Berichte

des Instituts für Massivbau der Leibniz Universität Hannover
Herausgeber:
Univ.-Prof. Dr.-Ing. Nabil A. Fouad
Leibniz Universität Hannover – Institut für Massivbau
Heft 18, Juni 2021

Fraunhofer IRB Verlag

Herausgeber
Univ.-Prof. Dr.-Ing. Nabil A. Fouad
Leibniz Universität Hannover
Institut für Massivbau
Appelstraße 9 A
30167 Hannover

ISBN (Print): 978-3-7388-0634-2
ISBN (E-Book): 978-3-7388-0636-6

Fraunhofer IRB Verlag
Fraunhofer-Informationszentrum Raum und Bau IRB
Postfach 80 04 60, 70504 Stuttgart
Telefon 0711 970-2500
Telefax 0711 970-2508
irb@irb.fraunhofer.de
www.baufachinformation.de
Druck: BoD – Books on Demand, Norderstedt

Chemical Prestressing of Thin Concrete Elements with Carbon Textile Reinforcement

von der
Fakultät für Bauingenieurwesen und Geodäsie
der Gottfried Wilhelm Leibniz Universität Hannover

zur Erlangung des Grades

DOKTOR-INGENIEUR

genehmigte

DISSERTATION

von

Katarzyna Zdanowicz, M.Sc.

2021

Tag der Promotion:	26.01.2021

Hauptreferent:	Prof. Dr.-Ing. Steffen Marx
Korreferent:	Prof. Ph.D. Dr.phil. Dr.techn. Konrad Bergmeister
Komissionsmitglied:	Prof. Dr.-Ing. Michael Haist
Vorsitz:	Prof. Dr.-Ing. Katharina Klemt-Albert

Summary

Prestressing of concrete with non-metallic reinforcement is an emerging field in structural engineering, as it can significantly contribute to better exploitation of the mechanical properties of such reinforcement and behaviour of concrete structures. Non-metallic reinforcement includes fibre reinforced polymer rebars and tendons, as well as various types of textile reinforcement. Textile reinforcement is particularly challenging as a material used for prestressing because of its geometrical form and structure. Therefore a technology of chemical prestressing, where expansive cement is used to induce stresses instead of mechanical devices, comes as an alternative yet promising method for such types of reinforcement.

In the thesis, the chemical prestressing technology is analyzed and applied to concrete elements with carbon textile reinforcement. The thesis aimed was to experimentally assess the behaviour of expansive concrete members with carbon textile reinforcement. Three objectives were undertaken: 1) to prove whether chemical prestressing of concrete elements with textile reinforcement is possible, 2) to quantify the introduced prestressing forces and the influence of prestressing on serviceability limit state with a focus on cracking load, 3) to analyse the influence of chemical prestressing on aspects relevant for the design of structural members and on their performance.

A feasibility study was performed to investigate the possibility of prestressing with expansive cement. Proper concrete ingredients were chosen and mixtures were designed, following by testing their material properties and the influence of expansive admixture in various amounts. Expansive strains were measured with standardized methods, as well as with distributed fibre optic sensors in specimens that were subsequently subjected to loading tests. Chemically prestressed concrete elements were tested for uniaxial tensile behaviour, in bond and flexural tests. Their behaviour was compared with this of non-prestressed ones.

Measurements of expansion in time and investigations of the behaviour of tested specimens have confirmed that chemical prestressing of elements with textile reinforcement can be realized. In specimens subjected to tensile tests, concrete prestress estimated from expansion measurements corresponded very well with calculations from tensile tests results, prestressing effect was also observed in beams with steel reinforcement and large scale slab specimens with textile reinforcement. Furthermore, specimens made of expansive concrete shown a significantly higher cracking load in comparison to reference specimens made of concrete without expansive admixture. At the same time, no deterioration of concrete material properties was observed, on the contrary, compressive and tensile strength increased with an addition of expansive admixture and the modulus of elasticity remained on the same level. Finally, an overall performance index was calculated which described the performance of chemically prestressed specimens compared to the reference members.

Zusammenfassung

Die Vorspannung des Betons mit nichtmetallischer Bewehrung ist ein Bereich des Bauingenieurwesens, der zunehmend an Bedeutung gewinnt, weil sie wesentlich zu einer besseren Ausnutzung der mechanischen Eigenschaften derartiger Bewehrung und des Verhaltens von Betonbauteilen beitragen kann. Zu den nichtmetallischen Bewehrungen gehören Faserverbundkunststoffstäbe und Litzen sowie verschiedene Arten von Textilbewehrung. Textilbewehrung ist als Material aufgrund der geometrischen Form und Struktur für die Vorspannung besonders herausfordernd. Aus diesem Grund stellt die Technologie der chemischen Vorspannung, bei der Quellzemente anstelle mechanischer Vorrichtungen zur Einleitung von Spannungen verwendet werden, eine alternative und vielversprechende Lösung für solche Bewehrungsarten dar.

In der Dissertation wird die Methode der chemischen Vorspannung analysiert und auf Betonbauteile mit Carbon-Textilbewehrung angewendet. Im Rahmen dieser Arbeit wurde das Verhalten von Quellbetonelementen mit Textilbewehrung experimentell erforscht. Es wurden drei Zielstellungen vorgenommen: 1) Nachweis, ob die chemische Vorspannung von Betonelementen mit Textilbewehrung möglich ist, 2) Quantifizierung der eingeleiteten Spannkräfte und des Einflusses der Vorspannung auf den Grenzzustand der Gebrauchstauglichkeit mit Schwerpunkt auf der Risslast, 3) Analyse des Einflusses der chemischen Vorspannung auf die für die Bemessung und Leistungsfähigkeit von Bauteilen relevanten Aspekte.

Zunächst wurde eine Machbarkeitsstudie durchgeführt, um die Möglichkeit der Vorspannung mit Quellzementen zu untersuchen. Die geeigneten Betonbestandteile wurden bestimmt und die Mischungen entworfen. Anschließend wurden ihre Materialeigenschaften und der Einfluss von Quellzusatzmitteln in verschiedenen Dosierungen untersucht. Die Dehnungen wurden sowohl mit genormten Verfahren als auch mit quasi-kontinuierlichen faseroptischen Sensoren in Probekörpern gemessen, die anschließend in Belastungsversuchen geprüft wurden. Chemisch vorgespannte Betonelemente wurden auf uniaxiales Zugverhalten, auf Verbund- und Biegeverhalten untersucht und mit den nicht vorgespannten Elementen verglichen.

Die Messungen der Dehnungen über die Zeit und die Untersuchungen des Verhaltens der Probekörper bestätigten, dass die chemische Vorspannung von Bauteilen mit Textilbewehrung erreichbar ist. Die aus Dehnungsmessungen geschätzte Betonvorspannung den Probekörpern entsprach sehr gut den Berechnungen aus den Ergebnissen der Zugversuchen, die Vorspannungswirkung wurde auch bei Balken mit Stahlbewehrung und großformatigen Plattenprobekörpern mit Textilbewehrung nachgewiesen. Darüber hinaus zeigten Probekörper aus Quellbeton eine deutlich höhere Belastung bei der ersten Rissbildung im Vergleich zu Referenzproben aus Beton ohne Quellzusatzmittel. Dabei wurden keine negativen Beeinträchtigungen der Materialeigenschaften des Betons beobachtet. Im Gegenteil, die Druck- und Zugfestigkeiten erhöhten sich bei Zugabe von Quellzusatzmitteln und das Elastizitätsmodul blieb unverändert. Schließlich wurde ein Gesamtleistungsindex ermittelt, der die Leistungen der chemisch vorgespannten Probe-körper im Vergleich zu den Referenzelementen darstellt.

Acknowledgements

As I came to Leibniz Universität Hannover in Germany as an exchange student during my master studies in civil engineering at the Cracow University of Technology in Poland, I joined a course led by Prof. Dr.-Ing. Steffen Marx. I still can clearly remember being impressed by his way of teaching and inspiring students. Thus, I am deeply grateful to have had Prof. Steffen Marx as a supervisor of my PhD thesis. I am thankful for the opportunity to work in the Institut für Massivbau, and especially grateful for his academic guidance along the way, time and eagerness to share his expert knowledge and for his sincere support in my professional and personal choices.

I would like to thank Prof. Dr. phil. Dr. techn. Konrad Bergmeister for his scientific interest in my work and for reviewing the thesis. My thanks also go to Prof. Dr.-Ing. Michael Haist for taking on the role of commission member and to Prof. Dr.-Ing. Katharina Klemt-Albert for chairing the examination.

Nothing can be achieved alone, and therefore I want to say thank you to all colleagues at the Institut für Massivbau. I am especially thankful for assistance with the experimental part to the laboratory technicians; René, Olaf, Hendrik, Ernst - Dankeschön! I want to express my great appreciation to Kerstin and Stephanie for their time and help with all administrative issues and to Boso for his genuine engagement. I want to particularly mention Steffen, Matthias and Chongjie who became great friends, thanks for all our coffee breaks together, our discussions and your support! I would like to give special thanks to all students who helped me with the daily work, especially Line, Dennis, Hubert and Tom. It was a pleasure to work at IfMa!

My grateful thanks are extended to all collaborative partners from the Max Bögl Bauservice GmbH & Co. KG Company. I highly appreciate the financial support provided by the Forschungsinitiative Zukunft Bau by the German Federal Institute for Research on Building, Urban Affairs and Spatial Development (SWD- 10.08.18.7-16.37) and the opportunities provided by the EU COST Action TU1207 "Next Generation Design Guidelines for Composites in

Construction" and the fib (International Federation for Structural Concrete) Task Group 5.1 "FRP Reinforcement for Concrete Structures".

Finally, I am deeply grateful to my parents for their support and - especially - to my husband Łukasz and our children for their lasting encouragement and understanding. Dziękuję Wam!

Contents

II EXPERIMENTAL RESEARCH 53

Symbols and Abbreviations

Abbreviations:

CFRP	Carbon Fibre Reinforced Polymer
COE	Coefficient of Efficiency
COV	Coefficient of Variation
CSA	Calcium sulfoaluminate
DEMEC	Demountable Mechanical Strain Gauge
DFOS	Distributed Fibre Optic Sensors
DIC	Digital Image Correlation
EA	Expansive Admixture
EPF	Elasto-Plastic Fracture model
FRC	Fibre Reinforced Concrete
FRP	Fibre Reinforced Polymer
LVDT	Linear Vertical Displacement Transducer
LWA	LightWeight Aggregate
OPC	Ordinary Portland Cement
PAE	PolyArylEther type superplasticizer
PCE	PolyCarboxylatEther type superplasticizer
RET	Restrained Expansion Test
RH	Relative Humidity
SAP	SuperAbsorbent Polymers

SCC	Self-Compacting Concrete
TRC	Textile Reinforced Concrete

Greek notation:

$\Delta\varepsilon$	strain residuum
ε_c	concrete strains
$\varepsilon_c r$	concrete strains at cracking
ε_s	steel strains
σ_{pre}	prestress value
$\sigma_{tu,max}$	maximum textile uniaxial tensile strength
$\sigma_{y,u}$	tensile strength of the yarn in textile reinforcement
ω_l	mechanical reinforcement ratio
τ_P	maximum bond strength

Latin notation:

A_{bond}	bond area of a textile roving
A_c	concrete cross-section area
A_t	steel reinforcement cross-section area
A_t	textile reinforcement cross-section area
b	width of an element
d	static effective depth of an element
E_c	concrete modulus of elasticity
E_s	steel reinforcement modulus of elasticity
E_t	textile reinforcement modulus of elasticity
f_{ctm}	direct tensile strength of concrete
$f_{ctm,fl}$	flexural tensile strength of concrete
f_t	textile reinforcement tensile strength
$F_{cr,(c,t)}$	force in concrete (c) or textile (t) at cracking
F_H	maximum pull-out load

h	height of an element
M_{cr}	cracking moment
$M_{Rd,est}$	estimated ultimate moment
N	normal force
w	crack opening

Cement shorthand notation:

A	Al_2O_3
C	CaO
C	CO_2
H	H_2O
M	MgO
S	SiO_2
$\bar{S}$	SO_3
T	TiO_2

Chapter 1

Introduction

1.1 Motivation

Prestressing of concrete elements is a way of optimizing their load-bearing capacity, durability, crack resistance and serviceability limit state, and it can be used to produce larger, thinner and more cost-effective concrete structures. Recently, besides fibre reinforced polymer (FRP) reinforcement, the topic of efficient application of concrete reinforced with carbon textile reinforcement (TRC, ger. *Textilbeton*) has gained increasing attention in Germany and worldwide (Triantafillou, 2016; Peled et al., 2017). TRC is sometimes described as a solution between unidirectional FRP rebars and fibre reinforced concrete (FRC) with randomly oriented short fibres. Its distinct feature is combining extraordinary mechanical properties of carbon or glass fibres with a dense structure providing sufficient ductility and deformability. At the same time, textile reinforcement has most often a form of a two-dimensional grid, which allows to reinforce elements in both directions simultaneously.

Prestressing of non-metallic reinforcement can contribute to better exploitation of its mechanical properties (Terrasi, 1998; Bergmeister, 1999; Schmidt et al., 2012; Zdanowicz, Kotynia and Marx, 2019). However, prestressing by means of conventional technologies which are applied for either steel or FRP reinforcement, is not easily adaptable for textile reinforcement. Firstly, carbon fibres are too sensitive to the transversal forces caused by mechanical anchorages and secondly, the structure and geometry of textile reinforcement do not allow to prestress all of the rovings simultaneously. First attempts to chemical prestressing of concrete with textile reinforcement proved that good results can be obtained, but such prestressing is not practically feasible and demands lots of effort (Reinhardt et al., 2003; Krüger, 2004). Actually, no reliable prestressing method was developed for textile reinforcement.

Therefore the concept of chemical prestressing, which involves expansive concrete instead of conventional mechanical anchorages and load cells to prestress the textile fibres, seems a promising solution to this problem. The chemical prestressing concept was invented in France in the '30s and further developed in Japan in the '60s, but besides numerous research activities and some pilot field applications did not come to engineering practice (Möll, 1954; Aroni et al., 1968; Klein et al., 1961; Okamura et al., 1977). Nowadays, this method might be particularly useful and feasible for textile reinforcement. However, to be able to apply the chemical prestressing method for TRC members, thorough experimental research, a deep understanding and reliable models of expansive concrete behaviour should be provided.

1.2 Research objectives

The aim of this thesis is to investigate the structural behaviour of chemically prestressed thin textile reinforced concrete members. Since chemical prestressing is a complex phenomenon, it is of utmost importance to initially design relevant concrete mixtures and additionally evaluate the strain development in time to be able to gain an adequate understanding of the behaviour of such prestressed elements. In the main part of this thesis, flexural, tensile and bond behaviour of chemically prestressed specimens were investigated because they are dominant in typical thin concrete constructions such as façade elements or slabs.

The specific objectives of the work presented in this thesis are:

1. To prove whether chemical prestressing of concrete members with carbon textile reinforcement is possible.

2. To quantify the introduced prestressing forces and the influence of prestressing on serviceability limit state, focusing on cracking load and pattern.

3. To analyse the influence of chemical prestressing on other aspects relevant for the design of structural members, that is on bond behaviour of the elements and material properties of concrete (compressive and tensile strength and modulus of elasticity).

1.3 Scope of the research and limitations

In the thesis, the flexural behaviour of chemically prestressed concrete members with steel and textile reinforcement is investigated. Firstly, strain development in time is analysed for various concrete mixtures; secondly, specimens are cast out of chosen concretes and finally bending, tensile and pull-out tests are conducted. During the tests the force-displacement relationship is investigated, strains are being measured in concrete and internal steel or textile reinforcement, and the cracking pattern is observed.

The topic of chemical prestressing of concrete members with textile reinforcement is a broad field with numerous remaining open questions and areas for further investigations. This thesis focused on the feasibility study and analysis of the structural behaviour of such elements. There is still plenty of limitations, which need to be considered, as they are relevant for a thorough understanding of the phenomena of chemical prestressing. The main limitations concerning the overall scope of work can be summarized as follows:

- long-term tests of creep and shrinkage were not carried out, yet they are necessary to characterize the structural behaviour during service life and possible prestress force losses,
- accelerated ageing of chemically prestressed specimens was also not a part of the thesis,
- investigations were carried out on the structural level, therefore no detailed material investigations are provided and changes in the chemical composition of concrete mixtures were not investigated here,
- the influence of environmental conditions on the chemically prestressed specimens was not studied,
- experimental sample sizes were restricted by the available resources.

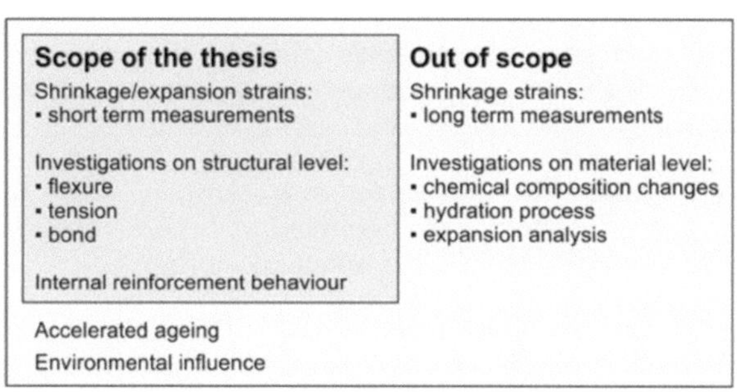

Figure 1.1: Scope of the thesis and relevant topics out of scope

1.4 Thesis outline

The first part "Theory" presents the current state of the art of knowledge regarding chemical prestressing with expansive admixtures, while the second part "Experimental research" includes methodology and results of own experimental research programme.

The following part summarizes the outline and contents of the thesis:

Chapter 1 – Introduction. Here the motivation, background information, research objectives, as well as the scope of the research and its limitations are provided.

PART I - THEORY

Chapter 2 – Chemical prestressing. Chapter describes the phenomena of chemical prestressing, its origins and history, as well as field applications and existing codes and standards.

Chapter 3 – Expansive materials. Expansive cements and admixtures are described here with a specific focus on CSA expansive admixtures and their compatibility with other concrete mixture ingredients. The influence of expansive admixture on concrete compressive strength, bending crack resistivity and creep is described. Furthermore, factors influencing expansion, the extent of reaction and re-expansion phenomena are discussed here and the state of the art in current knowledge about the mechanism of expansion is provided.

PART II - EXPERIMENTAL RESEARCH

Chapter 4 – Experimental programme. Chapter presents the concept of the experimental programme and conducted research as well as applied materials. Specimens, their casting and used nomenclature are described.

Chapter 5 – Concrete parameters. Various concrete mixtures tested within the work are addressed here as well as concrete compressive strength, modulus of elasticity and tensile strength of finally chosen concrete mixtures.

Chapter 6 – Expansion measurements. Selected strain measurement methods and their methodology are presented and the results of strain development in time are provided for all types of specimens.

Chapter 7 – Tensile and pull-out tests. Tensile tests on specimens with carbon textile reinforcement and their results regarding the effect of chemical prestressing are described and the initial study of pull-out behaviour of specimens with textile reinforcement is also addressed.

Chapter 8 – Flexural tests. Bending tests on beams with steel reinforcement and two types of slabs with carbon textile reinforcement are described and their results are presented. An analysis of prestress is included and discussed.

Chapter 9 – Conclusions. Summary of experimental research is given together with the discussion of the results and conclusions drawn from the investigations. Further research possibilities and outlook are described.

Part I

THEORY

Chapter 2

Chemical prestressing

2.1 Phenomenon of chemical prestressing

Chemical prestressing is a concept which assumes that stresses can be introduced into the internal reinforcement without any mechanical device but through controlled expansion of the concrete itself. In such a way tensile stresses are introduced into the reinforcement and compressive stresses in concrete (see Fig. 2.1). To achieve this effect three conditions have to be met:

1. an expansive admixture must be provided in the concrete mixture in a suitable amount – to cause expansion which is sufficient to exert stresses,
2. internal reinforcement, which serves as a restraint to concrete expansion, must be present - otherwise, free deformation occurs and no stresses are induced,
3. the expansion needs to take place within an effective period, when the bond between reinforcement and concrete is already provided but when the concrete is not yet too stiff, otherwise internal concrete structure might be damaged and material properties worsen.

By adjusting the amount of expansive admixture and taking into account the properties of the reinforcement as well as the curing conditions, chemically prestressed components can be designed and manufactured. This type of prestressing can be considered not only as an alternative way for unidirectional prestressing of composite reinforcement. For concrete elements with two- or three-dimensional textile reinforcement, there is also the possibility of uniform and self-acting prestressing in multiple directions.

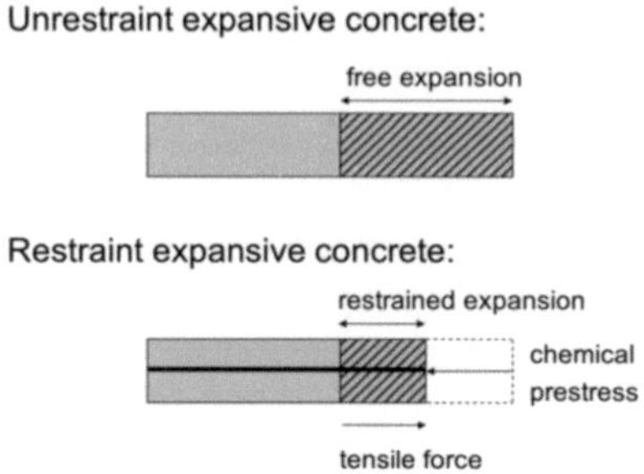

Figure 2.1: Schematic representation of chemical prestressing: free and restrained expansion

2.2 History

The idea of using expansive concrete to introduce prestress in concrete elements is attributed to Lossier, who started excessive works on the topic already in the 1930s in France (Billig, 1952; Möll, 1954; Magnel, 1956; Naaman, 2000). His expansive cement consisted of a mixture of ordinary Portland cement (OPC), sulfoaluminate cement and a stabilising compound, usually blast furnace slag. Lossier conducted experimental research on the material level, as well as model investigations on concrete decks, vaults and arches. He designed for example vaults with 50-cm long keystone made of expansive concrete (Fig. 2.2). Another application was an idea of a beam constructed of two separated parts, with continuous reinforcement going through both of them. The middle joint between two beams should be filled with expansive concrete and in such a way prestress can be applied (Fig. 2.3) (Möll, 1954).

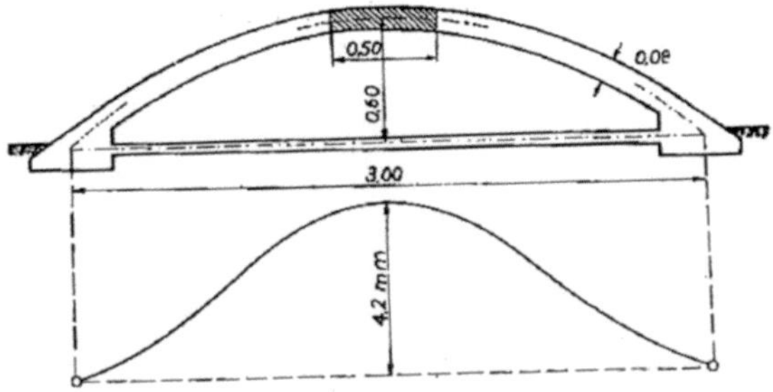

Figure 2.2: Vault with expansive concrete in keystone (Möll, 1954)

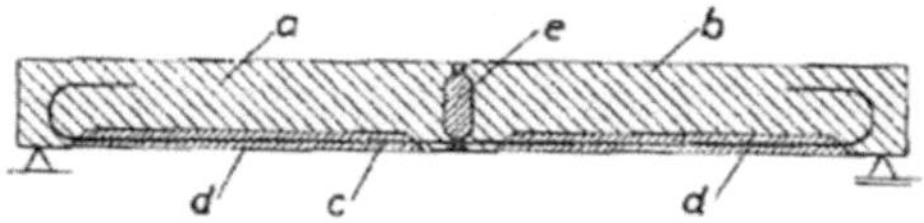

Figure 2.3: Beam with joint filled with expansive concrete (Möll, 1954)

Billig also summarized the developments of Lossier and concluded, that in 1952 it was "not yet possible to produce in this way preliminary stresses of specified value with any accuracy" (Billig, 1952). Despite this fact, expansive concretes were used to repair war-damaged arched bridges, as the application of expansive concrete in holes resulted in perfect bond as well as additional pressure on the arch.

The development of expansive concrete in its current form started again after II world war at the University of California at Berkeley, where Klein and Troxell developed an expansive compound, namely an anhydrous calcium sulfoaluminate (Klein et al., 1961). Later on, this compound was sometimes named "Klein compound" or "Type K cement" and is until now the basis for CSA admixtures type. First research activities and further developments of expansive concrete were pursued in following years (Lin and Klein, 1963; Aroni et al., 1968; Li, 1965; Mehta, 1967; Bertero, 1967; Benuska et al., 1971).

In Keil (1971), the development of expansive cements beginning from Lossiers' type until Type K cements are described in detail from the material point of view. In 1966 prestress with a magnitude of about 200 N/mm^2 in the steel reinforcement was introduced in laboratory conditions, which made the technology of chemical prestressing worth developing further to profit from practical applications. Krenkler (1980) described later developments in the expansive cements field and stated that the problem with expansive concretes is that they require loads of specific knowledge and therefore their industrial introduction is very slow although there is a need for such technology from the industry. Other researchers suggested however that "it is doubtful that chemically prestressed concrete will replace mechanical prestressing where high percentages of steel and high levels of prestress are required" (Aroni et al., 1968).

2.3 Field applications

Although numerous research studies and analyses can be found in the literature, not many field applications are described. Here findings of published applications of expansive cement are shortly summarized.

Expansive Cement in Japan

Nagataki (1980) summarized various applications and the market of expansive concretes in Japan up to the 1980s. In 1976 the production of expansive cement reached 1% of the total cement market. Generally, at this time in Japan concretes with expansive admixtures were divided into two groups: shrinkage compensating concrete (with expansive admixture content about 25 – 30 kg/m^3) and self-stressing concrete (40 – 60 kg/m^3). Numerous examples of application of expansive concretes could be found in Japan at this time – for instance, slabs on the ground (with the amount of expansive additive within the range of 26 – 40 kg/m^3) or underground structures, such as conduit tunnel linings (with increased amount of expansive admixture, up to 60 or 70 kg/m^3).

Furthermore, in Japan up to 1976, about 85% of the produced box culverts (i.e. 170 thousand tonnes) were chemically prestressed (Okamura et al., 1977). In 1995, about 1,700,000 m^3 of expansive concrete was used yearly with the share of 30% of cast-in-place concrete and 70% of precast concrete (Nagataki and Gomi, 1998).

According to a Technical Committee Report JCI-TC094A published by Japan Concrete Institute, nowadays, expansive concretes are still the most popular in Japan, where the ratio of use of expansive concrete to total ready-mix concrete increased from 0.4% to 1.6% from 1994 to 2006 (Hosoda, Morioka, Tanimura, Kanda, Sakai and Kishi, 2011).

Expansive cement in the USA

Numerous slabs of car parks, garages and highway pavements were cast already in the 1960s using Type K cement concrete, to avoid drying shrinkage cracking and/or compensate for high temperatures (Baltimore underground parking garage) (ACI Committee 223, 1970). Further, already in 1963 in Connecticut, an almost 500-meter long experimental section of pavement made of expansive concrete was cast besides numerous applications of shrinkage compensated concrete (Li, 1965). In 1972, an estimated 380,000 tonnes of expansive concrete were used in the USA (Hoff, 1972). This US Army report mentions, besides applications as shrinkage-compensating concrete, also that expansive concrete has the ability to prestress the reinforcement what "makes it very useful in precast operations as concrete pipes, building elements and modules, and tilt-up construction" (Hoff, 1972).

Besides, expansive concrete and its chemical prestressing possibilities were utilized by Naval Civil Engineering Laboratory for the application in thin-shell constructions (Keeton, 1973).

In the 1990s, Type K cement concrete was used for the construction of five bridge decks in Wisconsin. Besides these decks, comparison bridge decks were cast in conventional technology and the members were subsequently evaluated for cracking and surface distress. However, because of higher cost and unresolved "surface scaling issues" the application of expansive cement concrete was not proceeded later (Battaglia, 2012). On the other hand, similar projects were undertaken more recently in Georgia with promising results: reduced thickness, curling, cracking and fewer joints needed (Keith et al., 2006). The possible use of shrinkage compensating concrete reinforced with fibres for slabs on ground and concrete floors are also documented in a US Patent and should also lead to no shrinkage cracking, very good floor flatness and no detectable curling (Rice, 2003).

Field applications in Europe

Although the biggest market for expansive cements and admixtures are Japan and the USA, expansive cements are also already being applied in Europe. Three examples of field applications from Italy prove, that using expansive concrete mixtures with shrinkage-reducing admixtures (SRA) are a successful way to avoid shrinkage-related cracks or opening of joints. One of the construction sites in Italy where the combination of concrete including SRA and expansive admixture was used is the Museum of Arts of XXI century (MAXXI) in Rome, finished in 2006. Several concrete walls, 20 metres high and 60 metres long with sinuous shape should have no construction joints (Figure 2.4). Special self-compacting, expansive concrete was designed for this purpose. After its good performance, similar concretes were then successfully used for thick slab on the ground and a channel of a navigation lock (Troli and Collepardi, 2011).

Figure 2.4: Joint-less walls of the MAXXI in Rome (Troli and Collepardi, 2011)

In Germany, the usage of expansive cements based on CSA was investigated and analysed in the 1990s (Breitenbücher et al., 1992), however, no more field applications can be found in the literature. Expansive concretes are currently mainly used in concrete slabs on the ground with steel fibre reinforcement. Application of CSA admixtures in such elements can compensate tensile stresses due to restrained contraction and reduce curling et the edges. Recently a concept of joint-free slabs on the ground made of expansive concrete with steel fibres has been described in (Destrée et al., 2018), with several field applications in Denmark, Lithuania and Norway. In general, curling-free slabs and screeds is a very important reason for applying shrinkage-free concrete with CSA admixtures (Péra and Ambroise, 2004).

2.4 Codes and guidelines

Currently, there are two guidelines or standards which consider expansive concrete in general. First in the USA, ACI 223R-10 "Guide for the Use of Shrinkage Compensating Concrete" was published in 1998 and actualised in 2010 (American Concrete Institute, 2010). The second guideline exists in Japan since 1993, JSCE No. 75 "Recommended praxis for expansive concrete" (Japan Society of Civil Engineering, 1994).

ACI 223R-10 guide deals however only with shrinkage compensating concrete and not with chemical prestressing technology. Nevertheless, it considers various restraint conditions, reinforcement ratios and arrangements and can therefore be a useful source also for expansive concrete applications where expansion is higher than only to compensate for shrinkage. Expansive cements which are available in the USA are described more in detail in Chapter 3.1. The guideline gives provisions for structural members such as structural RC slabs, slabs-on-ground and their joints, post-tensioned elements and walls. It provides information about concrete mixture compositions and compatibility with other admixtures. For RC slabs, ACI Guideline states that although the flexural strength of shrinkage-compensating concrete remained uninfluenced, there is an effect on the load at which flexural cracking occurs - cracking load is higher in shrinkage-compensating concretes in comparison with portland-cement concretes.

JSCE Guidelines (Japan Society of Civil Engineering, 1994) on the other hand from the beginning differentiates between shrinkage compensating and expansive concrete (Fig. 2.5). Expansion of concrete at the early stage (2 or 3 days) should be measured according to JIS A 6202 standard (JIS, 1999), then measurements should be repeated at 7 days and continued further. At 7 days, if expansion values are within the range of 0.15 mm/m and 0.20 mm/m, concrete can be designated as shrinkage compensating. When expansion at 7

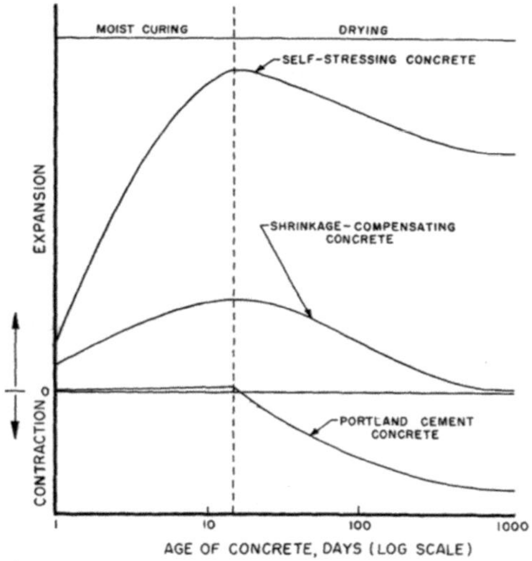

Figure 2.5: Shrinkage compensating and self-stressing concrete in comparison with portland cement concrete (Aroni et al., 1968)

days is not lower than 0.20 mm/m and not higher than 1.0 mm/m, concrete is called chemically prestressed. Expansive admixture contents should be determined by testing to obtain the required expansion ratio, however according to JSCE content equal to 30 kg/m^3 can induce expansion rates between 0.200 – 0.700 mm/m, whereas with content up to 65 kg/m^3 values around 1 mm/m can be obtained.

When shrinkage compensating concrete is used, so the expansion rate is under 0.20 mm/m, the influence of induced stresses on existing reinforcement does not need to be considered. When higher values of expansion rates are achieved and expansion is restrained for example through internal reinforcement in concrete, chemical prestress will be induced – which will cause improvements in crack strength and displacement and deformation properties. Such members should be considered as prestressed concrete elements. However, the influence of the expansion of concrete on the reinforcement should be taken into consideration only during calculations of axial strength, and not for flexural tensile strength. It is explained that chemical prestressing does not have a high influence on the lever arm length and thus on the ultimate limit state.

Chapter 3

Expansive Materials

3.1 Expansive cements and admixtures

Expansive cement is in general, according to the ACI 223R-10, "a cement that, when mixed with water, produces a paste that, after setting, increases in volume to a significantly greater degree than does portland cement paste; used to compensate for volume decrease due to shrinkage or to induce tensile stress in reinforcement" (American Concrete Institute, 2010). There are various types of premixed expansive cement which in the United States of America are differentiated by four letters:

- Type K - a mixture of portland cement, anhydrous tetracalcium trialuminate sulfate (C_4A_3S), calcium sulfate ($CaSO_4$), and lime (CaO),
- Type M - interground or blended mixtures of portland cement, calcium-aluminate cement, and calcium sulfate,
- Type G - where the primary chemical mechanism for expansion is the reaction of calcium oxide and water to create the expansive calcium hydroxide,
- Type S - a portland cement containing a high computed tricalcium aluminate (C3A) content and an amount of calcium sulfate above the usual amount found in portland cement.

The difference between types S, M and K is presented in Figure 3.1. Generally speaking, Type K cement was developed in the USA, while Type M cement in Russia (Krenkler, 1980). All these cements do not require further mixing with portland cement, however, it is therefore also not possible to adjust proportions in order to achieve the desired expansion. In Russia, expansive cements are additionally divided into three classes: 20, 40 and 60 and they provide prestress in a magnitude of 2, 4 or 6 N/mm^2 (Kurdowski, 2014).

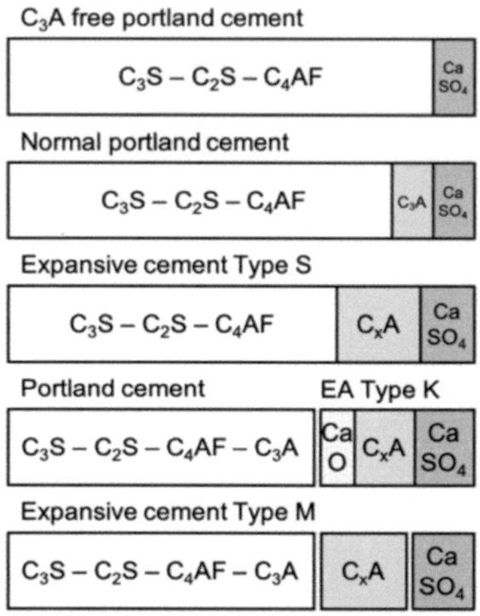

Figure 3.1: Expansive cement types S, M and K compared with normal and C_3A free portland cement (based on (Krenkler, 1980))

In Europe and Asia, this nomenclature is less widely used. The most common expansive cements are so-called calcium sulfoaluminate cements (CSA), which are in fact comparable with Type K cements. They contain typically between 30–70 mass % ye'elimite (C_4A_3S), belite, gehlenite and calcium aluminates (Winnefeld et al., 2017). Which phases are formed during hydration of CSA cements depends also on the calcium sulfates which are added. In the presence of calcium sulfates (i.e. anhydrite, hemihydrate or gypsum), ettringite and aluminium hydroxide are formed. In absence of calcium sulfates, when only water is available, the process of hydration of ye'elimite results in the formation of monosulfate and aluminium hydroxide. Also, the presence of lime in various form is attributed to differences in hydration processes (Trauchessec et al., 2015). Expansion mechanisms are complex phenomena that are believed to be linked with ettringite formation, although not yet thoroughly understood. The previous and currently viable theories of expansion are presented in chapter 3.7.

Expansive properties are not the only difference between CSA and OPC cements. CSA cements have also shorter setting and hardening times than OPC, which makes them good for repair purposes, but in casual applications often demand using retarders (Pimraksa and Chindaprasirt, 2018). The compatibility of CSA cements with commonly used retarders is described in chapter 3.2.1. Furthermore, one of their drawbacks is a lower pH after hydration, what can cause faster corrosion of steel reinforcement. Lower alkalinity is however not relevant when non-corrosive, composite or textile reinforcement is used or can be even advantageous for reinforcement made of glass fibres (Péra and Ambroise, 2004).

CSA cements are becoming more recognition recently also because they are a potential low carbon dioxide alternative to ordinary Portland cements (?Hanein et al., 2018). Lower emissions of CO_2 are connected to two reasons: firstly, the clinkerization temperature is around only 1250°C (which is 200°C lower than for OPC clinker) and, secondly, the percentage of limestone is lower (Winnefeld and Lothenbach, 2010; Trauchessec et al., 2015). In comparison of OPC with CSA cements, not only is the CO_2 emission lower (-43% for calcium sulphoaluminate cement), but also the specific heat consumption during clinkering (-14% lower) and crushing energy (-40% lower) (Justnes, 2007). Furthermore, concrete with CSA admixture is being investigated as self-healing concrete, as its expansive properties allow to effectively close cracks with the width of 250–400 µm and to decrease the permeability of concrete (Sisomphon et al., 2012). Last but not least, CSA cements are also studied regarding their potential to stabilize radioactive wastes (e.g. borate-containing), as their rate of hydration is less retarded by borate ions than that of ordinary Portland cements (Champenois et al., 2015).

Currently, more commonly used are so-called expansive admixtures which can be intermixed with Portland cement in desired ratio and proportions. Expansive admixtures are generally divided into basing on ettringite formation (mainly calcium sulphoaluminates) and portlandite formation (mainly based on magnesium oxide). Figure 3.2 shows a three-component diagram of expansive admixtures relating to calcium sulfoaluminate series. When concrete hydrates, expansive hydrates are being produced: ettringite (T), monosulfate (M) and calcium hydroxide, CaO (Nagataki and Gomi, 1998).

In the work in this thesis, DENKA CSA #20, one of the commercially available CSA-based admixtures was used. DENKA CSA #20 is an admixture which contains mostly calcium-aluminium-sulphate ($3CaO \cdot 3Al_2O_3 \cdot CaSO_4$), Anhydrite (Calciumsulphate, $CaSO_4$), and Calcium oxide (CaO).

3.2 Factors influencing expansion

The expansion of concrete with CSA admixture can be influenced by various factors. In complex compositions, such as concrete mixtures nowadays, the number of issues that should be considered is high. Eventually, the expansive

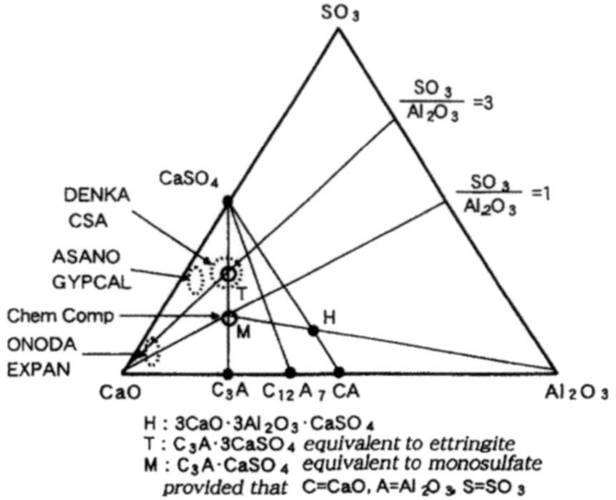

Figure 3.2: Three-component diagram of expansive admixture (Nagataki and Gomi, 1998)

behaviour of a new concrete mix must be tested before desired application. Nevertheless, there is a bundle of recommendations and results of previous investigations, which can help in the design of a proper mixture. Here the most important for the experimental part of this thesis are summarized.

3.2.1 Compatibility with other materials

In general, CSA admixtures can be mixed with Portland or other types of cements, however particular concrete mixture ingredients can interfere with them and influence the expansion or compressive strength of concrete. ACI 223R-10 states, that accelerating admixtures based on calcium chloride should not be used with expansive admixtures, and for all other accelerators, air-entraining, water-reducing or retarding admixtures a trial batch and a restraint expansion test should be conducted to determine their effect on expansion (American Concrete Institute, 2010). Numerous further recommendations regarding the compatibility of CSAs are available in the literature, several most relevant are described in the following sections.

Cements

In general, cements that are not rapid-hardening should be used to let some time for initial expansion before achieving full strength. A study by Ito et al. (2004) shows that concrete mixtures containing low-heat cement with expansive admixture or low-heat cement with expansive admixture and shrinkage-reducing agent show higher expansion than similar mixtures with ordinary Portland cement.

This finding was later confirmed by research, which checked the influence of various cement types on the chemical prestress effectiveness (Iijima et al., 2013). Two amounts of expansive admixture were tested, i.e. 20 kg/m^3 and 40 kg/m^3. Cements used in this research were: ordinary portland cement, low-heat portland cement and high early strength portland cement. The conclusion was that for higher dosages of expansive admixtures, which should introduce not only shrinkage reduction but also chemical prestress, the use of low-heat cement can be desirable as the prestrain achieved for these specimens was almost twice as high as for specimens made of ordinary cement (Iijima et al., 2013). Furthermore, high calcium aluminate cements which are prone to a rapid decrease of compressive strength should rather not be used as a component of expansive concrete (Nocuń-Wczelik et al., 2011).

Retarders

As CSA cements are generally rapidly setting and hardening, the use of retarders is often unavoidable to achieve proper workability. However, the commonly available retarders also influence the hydration process of CSA (Zajac et al., 2016; Hu et al., 2017). The usage of borax and citric acid retarders influenced the setting times of the calcium sulfoaluminate cements: when the amount of retarder increased, also the initial and final setting time increased (from 20-30 to about 55-65 minutes). Although the 1-day compressive strength of CSA cement decreased when borax or citric acid were used, the decrease was not observed any more after 28 days, and for mixtures where retarder was used in an amount of 0.5 - 0.6% by weight of cement, even a strength increase was observed (Hu et al., 2017).

The effect of Na/K-tartrate, Na-gluconate and borax retarders on the early age hydration of CSA cements was also investigated at a single addition rate (2% by mass of cement) and it was proved, that all three retarders suppressed or slowed down the initial formation of ettringite (Zajac et al., 2016). The differences between those retarders were that the addition of tartrate and gluconate inhibited the growth of ettringite (gluconate slows down the reaction more than tartrate) and increased the pH of cement paste, while the use of borax inhibits dissolution of ye'elimite and lowers the pH. Finally, on scanning electron microscopy (SEM) it was observed, that the final morphology of ettringite also depends on the retarder type (Zajac et al., 2016). There is however no direct linkage between the type of retarder used and the expansion of CSA cement.

Silica fume and fly ash

Silica fume is said to reduce the expansion of CSA-based cements, probably because the pH of pore solution is lowered and in presence of silica fume, there is not enough water for hydration (Lobo and Cohen, 1992). For fly ashes, the topic is more complex. Addition of fly ashes in general decrease the chloride permeability of expansive cement mortar mixture (Folliard et al., 1994). According to (Chaunsali and Mondal, 2015*b*, 2016), an addition of class C also reduces expansion and shortens the period when expansion takes place, whereas class F has a positive effect and increases expansion. However, Garcia Calvo et al. (2017) used class F low CaO fly ashes and Type K expansive agent in a self-compacting concrete mixture and concluded that the expansion performance was worse than in plain SCC concrete without fly ashes. The explanation is that $Ca(OH)_2$ is consumed by hydration of fly ashes or that setting time is prolonged. Limestone fillers are recommended instead.

Limestone

Free lime, or CaO, has an advantageous effect on expansion (Kurdowski and Thiel, 1981; Polivka, 1973). Previously, it was believed that in absence of lime, no effective expansion takes place (Mehta, 1973). This theory was denied for example by Mather (1984), and although not necessary, lime is indeed important for effective expansion. The contribution of limestone to the hydration of calcium sulfoaluminate cement was studied recently by (Martin et al., 2015), who by calorimetric analysis concluded that limestone can accelerate the hydration reaction. The addition of limestone to the OPC with CSA admixture is not only economically advantageous but also indirectly stabilizes ettringite and increases the volume of formed hydrate phases and enhances the compressive strength development. This stabilising effect (based on the creation of monocarbonate instead of monosulfate) for OPC was also described previously (Lothenbach et al., 2008). The presence of lime is also said to retard the hydration of ye'elimite ($C4A3\hat{S}$) during the first 6 hours (Chaunsali and Mondal, 2015*b*).

Gypsum

The presence of gypsum also positively influences the hydration of CSA cement. With addition of gypsum, hydration is faster (according to isothermal calorimetry results) than with anhydrite only (Winnefeld et al., 2017). Improvement in strength development and hydration kinetics is visible already when 20% of anhydrite is replaced by gypsum. CSA cements that contain less than 30% gypsum by mass are said to be rather non-expansive, while a high expansion up to 4 mm/m in air drying after 14 days was observed for a 50% CSA clinker–50% gypsum mixture (Telesca et al., 2014).

Aggregates

Expansion can differ also when particular aggregates are used. As mentioned above, expansion increases when aggregates contain gypsum or other sulphates but decreases when large amounts of chlorides are provided in aggregates (American Concrete Institute, 2010). Furthermore, aggregate types which can accumulate water in their pores (e.g. lightweight aggregates, LWA) can contribute to higher expansion rates (Aroni et al., 1968). Besides LWA, also superabsorbent polymers (SAP) can be used to increase the expansion (Wyrzykowski et al., 2018). Finally, according to (Mailvaganam, 1995), aggregates of high elastic modulus contribute to larger expansion. Therefore a recommendation is made to use the same aggregates during initial expansion tests that will be used in actual applications.

Self-compacting concrete

In general, because of the higher volume of paste, self-compacting concretes (SCC) tend to have higher shrinkage than conventionally vibrated concrete (Loser and Leemann, 2009; Leemann and Lura, 2014). Therefore, the use of expansive admixtures in SCC mixtures was described as even more advantageous than in normal concrete, as in SCC "a significant improvement in post-cracking can be expected" (Kishi et al., 2000). Even in poorly cured SCC beams, cracking load and ductility were improved. (Garcia Calvo et al., 2017) compared conventionally vibrated concrete with self-compacting concretes either with limestone filler or fly ashes and two types of expansive admixtures, Type K and Type G. They observed no significant differences in expansion behaviour between conventional and SCC mixtures. The most remarkable observation was that in the SCC mixes air content measured was higher (Garcia Calvo et al., 2017).

3.2.2 Mixing time and temperature

Mixing water should be possibly free from sulphates or chlorides, otherwise, it can the expansive properties of concrete (Mailvaganam, 1995). Another important factor is the mixing time. If mixing is unnecessarily prolonged, more ettringite is formed during mixing and so the rate of expansion later on decreases. Especially at higher temperatures prolonged mixing is a serious threat (Polivka, 1973). Also when the temperature is too low, the expansion will be inhibited and final strains will decrease (50% decrease for concrete with the same composition but cured at 7°C in comparison to 21°C) (Polivka, 1973). However, too high temperatures are also not advantageous, as they cause the strength to increase more quickly. So although the expansion rate is higher at the beginning, the final strains might be lower than for concrete cured at normal temperature.

3.2.3 Geometry and restraint

Expansion occurring at the early stage, when concrete has not yet gained enough strength and the bond between concrete and reinforcement is weak, has a much lower contribution to total prestress amount than expansion occurring later. Thus proper curing of concrete members, which provides water for later stages of hydration, is of great importance. Therefore, also the influence of dimensions of concrete members needs to be taken into consideration – elements with higher depths have different moisture contents along their depth, due to the fact that external moisture curing can not penetrate the whole depth to the same extent. Thus, expansion values can differ within the element (Aroni et al., 1968).

Not only is the geometry important, but also the restraint is considered the most significant issue for chemically prestressed concrete. Uni-, bi- or triaxial restraint can be applied for a concrete member, among which internal, constant restraint by reinforcement can be distinguished from external restraint, which can be temporary – e.g. to encounter only for the expansion occurring in the first phase of curing of the element. Reinforcement embedded in concrete needs to be analysed during the design of chemically prestressed members. Also during the construction phase, unpurposive changes of the location of rebars can induce additional, unplanned stresses and unsdesirable distortions.

Surface elements need only biaxial restraint and their reinforcement ratio is usually of moderate values, therefore chemical prestressing can be a particularly good solution for such members. In prismatic structures such as beams or columns usually higher prestressing stresses are necessary and not always they can be achieved through expansive concretes. According to (ACI Committee 223, 1970), chemically prestressed concretes might also require mild bi- or triaxial restraint because lateral deformations due to expansion might have a detrimental effect on the concrete strength. Furthermore, restraint should be symmetrical, otherwise differences in deformation and subsequent warping might occur.

However, in most of the research published, only uniaxial reinforcement is provided as internal restraint, the same is valid for most of the theoretical models for prestress analyses. Results with biaxial restrain are scarce (Okada et al., 1983; Ishikawa and Tanabe, 2010). It is believed that when specimens have biaxial restraint, the chemical prestressing in one direction reduces the expansion in another direction. Okada et al. (1983) introduced a formula (Eq. 3.1) which attributes for this decrease:

$$\varepsilon_{RX} = \varepsilon_{Rx0} - K_x \cdot \varepsilon_{RY} + \nu \cdot \varepsilon_{cey} \tag{3.1}$$

where:

ε_{RX}, ε_{RY} - restrained expansive strains under biaxial restraint in the direction of X and Y,
ε_{Rx0} - restrained expansive strains under uniaxial restraint,
ε_{cey} - elastic strain caused by chemical prestressing,
K_x - expansion reducing coefficient,
ν- Poisson's ratio.

If no restraint is provided, microcracks occur in expansive concrete specimens (Sakai and Kishi, 2012). Figure 3.3 shows a result of SEM analysis of two unrestraint specimens, made of normal concrete (left) and concrete with expansive admixture (right) with a magnification of 300 times. The length of microcracks is around 10µm in normal concrete and around 100µm in expansive concrete. Such microcracks in concrete structure might cause nonlinear behaviour of expansive concrete when compared with normal concrete (Sakai and Kishi, 2012). When restraint is given the changes in volume are limited and expansion strains are lower, which decreases the development of microcracks.

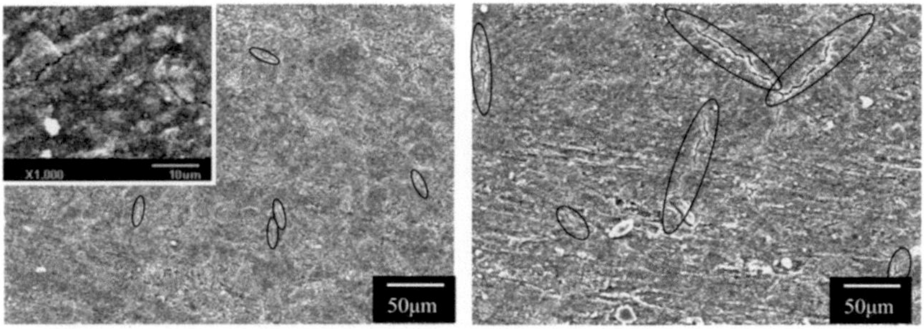

Figure 3.3: Microcracks visible in normal concrete (left) and expansive concrete (right) in SEM pictures with magnification of 300 times (Sakai and Kishi, 2012)

3.2.4 Water curing and re-expansion phenomena

As mentioned above, expansion is strongly influenced by water curing, either from an external or internal source. A sufficient water supply is essential to obtain a proper expansion of concrete. Also, by limiting the amount of water or the time when it is available, the expansion degree can be controlled.

In general, curing in water is often performed and considered as advantageous, although only in the early age of concrete – according to (Okamura et al., 1977), up to maximum of 10 days after concreting. Although curing in water provides the most rapid expansion, the same expansion rate can be obtained with curing in fog (loose-fitting membrane) (Klein et al., 1961). Hoff (1972) indicates, that water curing for minimum of three days is optimal, although curing membrane can also be applied. It is important to retain all the moisture in concrete and prevent evaporation, therefore curing with plastic sheets had limited success (Hoff, 1972). Another possibility of providing enough water supply is the method of internal water curing. There are two main methods of internal water curing – using pre-saturated lightweight aggregates or super-absorbent polymers (José Oliveira et al., 2014). These methods can contribute to the mitigation of shrinkage and may function as well with expansive admixtures.

The early-age curing is indispensable for expansion and if in the first period the members are not cured properly, expansion can take place later when members are exposed to rain or high humidity atmosphere. According to the JCI Technical Committee on Performance Evaluation of High Performance Expansive Concrete (Hosoda, Minoru, Tanimura, Kanda, Sakai and Kishi, 2011) when the water/binder ratio is lower than 0.3, expansive admixture might remain unreacted and delayed re-expansion can take place. However, if an expansive admixture with higher fineness is used, its efficiency can increase so that there is no unreacted part remaining.

To understand the re-expansion phenomena, it is necessary to ensure which part of the expansive admixture added to the mixture has already reacted. If unreacted expansive admixture is present and water is provided later (eg. in form of rain, humidity or external supply), expansion may continue. The extent of reaction of ettringite can be described with a parameter R, which is a ratio of supplied $C\overline{S}$ to required $C\overline{S}$ (Sousa Ribeiro, 1998). It is, in other words, "percentage of reactivity of expansive agent in expansive cement blend (the supplier of aluminates) controlled by means of sulphates". R is a parameter that reflects the extent to which the substrates convert to ettringite. If $R = 100\%$, the reaction was the most efficient and the whole expansive agent reacted, however sometimes in this case the magnitude of expansion is too high and leads to the destruction of the material. On the contrary, values of R lower than $R = 55\%$ mean negligible and ineffective expansion (Sousa Ribeiro, 1998; Olek and Cohen, 1991).

Bertero (1967) investigated the re-expansion phenomena. The research included ten specimens, out of which four were placed into water curing with 1-day delay (after casting), and the other have following delays in begin of water curing: 3, 6, 8, 10, and two specimens delayed by 11 days. After casting and before placing in water, specimens were wrapped in polyethylene sheets and stored at room temperature. It was observed, that in the time of first 24 hours no expansion occurred for specimens without water curing, and the expansion was also delayed for specimens that were without water curing for a longer time. The expansion was retarded as soon as specimens were placed in water again, the longer was the delay period, the higher expansion rate was observed within the first four days and the longer was also the time necessary to achieve a stabilised value of expansion. However, the total loss of expansion was not higher than 10% and the compressive strength was also not influenced, contrary to the stiffness of the specimens. Secant modulus of elasticity was approximately 30% lower than for conventional member (Bertero, 1967).

Thus, understanding the mechanism of expansion, and possibly the ability to measure or calculate the extent of expansion is necessary to avoid the undesired increase in introduced prestressing.

3.3 Influence on compressive strength

As far as the changes in compressive strength are considered, two schools of thought are present in the literature. The first one states that a decrease in compressive strength is unavoidable when expansive agents are used and it is directly related to the amount of expansive admixture. The second one claims that strength decrease is not necessarily to be observed, either because changes in the structure of concrete influence compressive strength positively or because reinforcement provides sufficient restraint for concrete.

And so, Klein et al. (1961) claim that when expansion exceeds 0.3% - 0.4%, the compressive strength decrease is large and achieves up to 50%. Also, Bertero (1967) investigated the influence of curing time on the compressive strength and observed that for expansive concretes compressive strength increases only within the first ca. 10-12 days of water curing. Afterwards compressive strength values decreased and compressive strength of specimens cured under water for 21 or 28 days was about 40% lower than for specimens cured for 5 days. Such a significant drop in strength was justified by an increase in mortar stiffness and transverse expansion ratio. The transverse direction was unrestrained (transversal expansion reached up to 2%, while longitudinal about 0.3%) and therefore additional strains caused deterioration of concrete mechanical properties.

Another theory says that compressive strength might increase because as expansion proceeds, concrete "heals itself" and fills in the pores and empty spaces in its structure (Aroni et al., 1968). In such a way compressive strength of expansive concretes might be higher than that of normal concrete. Mailvaganam (1995) defines an expansion threshold - when expansion exceeds 0.3% and is unrestrained, the strength of concrete can be reduced, but when enough restraint is guaranteed the decrease of compressive strength can be avoided.

Similarly, Jia et al. (2016) present a hypothesis that up to some extent, the concrete structure gets denser (as the pores are filled with expansion products), and above this threshold, expansion leads to defects in the structure of concrete and therefore a decrease in strength is observed. A strength increase of 2% and a decrease of 15% was observed with expansive admixture amount in the range between 7% and 25%. On the other hand, experiments conducted on Type K expansive admixtures showed almost no influence (less than 1% loss) on compressive strength, although the additive was applied in high amounts reaching 15% and 20% (Carballosa et al., 2015).

The positive influence of expansive admixtures on compressive strength was also observed when expansive admixture was used simultaneously with SRA (José Oliveira et al., 2014). The use of common SRAs caused a decrease in compressive strength of concrete equal to around 30%, and this decrease remained constant after 28 days (experiments were repeated after 9 months). However, when SRA was used simultaneously with expansive admixture, a lower reduction of compressive strength was observed after 28 days (up to 23%). What is more, strength developed further after 28 days and therefore a kind of recovery was observed as after the next 9 months there was almost no reduction in compressive strength (up to 1.6% decrease).

In conclusion, there is no unified opinion of the researchers on the influence of CSA-based expansive admixtures on the compressive strength of concrete. Results obtained within this work are presented in Chapter 5.1.

3.4 Influence on bending crack resistivity

Bending crack resistivity, or the load at which cracks occur when specimens are subjected to flexure, might be also influenced by the use of expansive admixtures. According to Okamura et al. (1977), the bending crack resistivity is increased by using expansive concrete and producing chemically prestressed elements. Furthermore, bending crack width of chemically prestressed concrete members should be smaller than in non-prestressed elements. The same is valid for diagonal crack resistance when shear forces are acting – when chemical prestress of concrete between 2 and 5 N/mm^2 was introduced into elements, their diagonal crack resistance was increased by 20 to 40% in comparison to non-prestressed concrete (Okamura et al., 1977).

Hosoda and Kishi (2001) claim that improved cracking load and reduced crack width are considerable advantages of chemical prestressing. However, they state that there is a difference in the tensile behaviour of conventionally prestressed concrete members and elements prestressed chemically. In tensile tests, conventional specimens exhibit linear behaviour up to cracking, whereas chemically prestressed ones show nonlinear behaviour (i.e. plastic deformability) even before cracking (Fig. 3.4). This leads to a conclusion, that the crack resistance is improved not only due to prestress, but also because deformability is greater. It is believed that plasticity is a result of the collapse of voids in concrete and of slippage between aggregate and mortar. Further, the contribution of plasticity in expansive concrete to the crack resistance is so high, that "the effects of expansive agent cannot be rationally evaluated only in terms of stress" (Hosoda and Kishi, 2001).

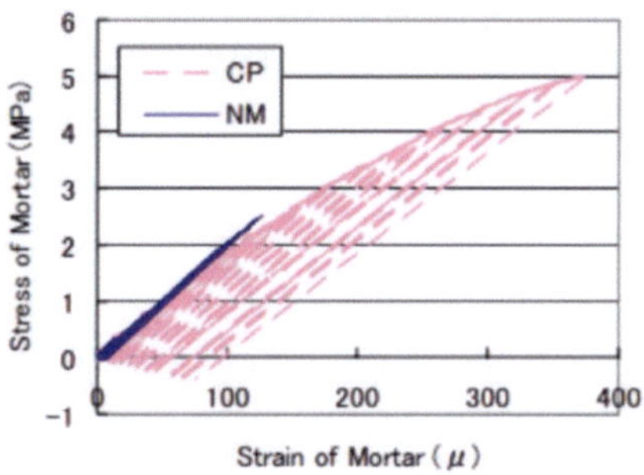

Figure 3.4: Stress-strain relationship under tensile test of normal mortar (NM) and expansive mortar (CP) after 3 days (Hosoda and Kishi, 2001)

Elasto-plastic fracture (EPF) model is adopted which for uncracked concrete classifies sources of nonlinearity into fracture and plasticity, where the fracture is understood as the "elasticity damage caused by dispersed defects with no clear localization" and plasticity means unrecoverable deformation (Fig. 3.5) (Maekawa et al., 2003). Thus, a fracture affects ultimate strength while plasticity the deformation of concrete.

Concrete is modelled as a set of elasto-plastic components, where springs refer to elasticity and sliders to plasticity. Breaking those springs means that fracturing damage occurs. For chemically prestressed concrete, in the EPF model we can assume that new elements (spring-slider pairs) are produced during the hydration process, and due to expansion, each element has its initial strain even before the tensile strain is applied (Hosoda and Kishi, 2001). Therefore, during loading various spring-slider pairs are damaged at different load levels, which leads to observed stiffness decrease. Fracture is caused by accumulated damage in concrete structure: microcracks in the structure, microscopic buckling and collapse of the mortar and aggregates (Hosoda and Kishi, 2001). EPF model explains why for expansive mortar a stiffness decrease is observed after unloading when maximum tensile stress increases, while for normal mortar there is no difference in stiffness.

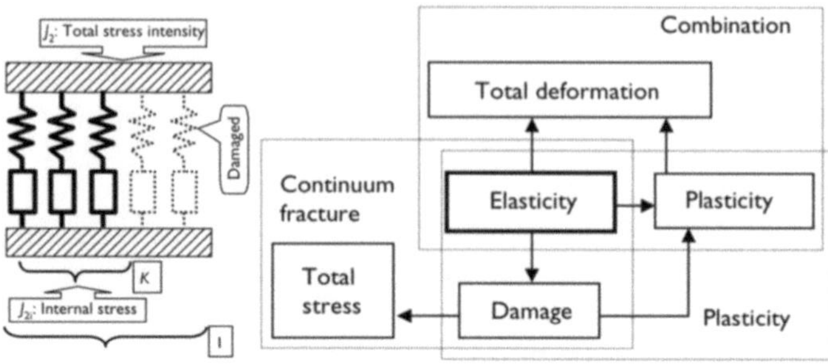

Figure 3.5: Scheme of the elasto-plastic fracture (EPF) model (Maekawa et al., 2003)

Hosoda and Kishi (2001) claim also, that expansive strains become uniform in sections more than 500 mm from the member's end, therefore in four-point bending tests, a sufficient length is crucial to provide uniform moment distribution in the designated area. This issue is further addressed in chapter 6 with expansive strain measurements.

3.5 Creep of expansive concrete

Creep deformations belong to the load and time-dependent deformations that concrete undergoes from the beginning of its loading and that continue to progress during its use. Creep deformations must be separated from shrinkage and expansion deformations since these two components are load-independent. The creep of concrete results mainly from the creep tendency of the hardened cement paste. On the contrary, the creep of aggregates is relatively low and can usually be neglected.

From the field of prestressed concrete structures, it is known that concrete creeps under compressive stress and that it leads to losses of prestress force. Regarding the creep of expansive concrete, there is an open question if the advantages of expansive concrete will not be utilized fully because of creep of concrete, which will eventually decrease the prestressing effect. Unlike the investigations of the shrinkage behaviour, there are still not many studies on the creep behaviour of the chemically prestressed concrete.

Okamura et al. (1977) assumed that lower creep deformations occur when CSA is used. They support their assumption with the fact that since the expansion of concrete begins at a very early age and compressive stresses which are introduced into concrete increase gradually, the prestress has already been influenced by creep up to that time. This implies that following creep deformations will be lower than when prestress is mechanically applied "all at once". Furthermore, it is believed that expansion energy exists in concrete elements even after the time when expansion seems finished. The reason for such belief is because although the expansion is constant after 28 days under restrained conditions when restraint is temporarily released, some more expansion can be observed even after a longer time. It can contribute to contradicting creep deformations (Okamura et al., 1977).

Aroni et al. (1968) also argue similarly. The creep deformations are lower if expansion admixtures are added to the concrete. This is attributed to the creep process, which already takes place when the expansion agent is added as a result of the expansion and the associated self-prestressing. Creep under external load is therefore substantially reduced since part of the creep deformation has already been completed.

A recent study by Sirtoli et al. (2019) investigated the influence of CSA cement on creep behaviour. Three different mixtures were used: pure Portland cement, complete replacement of the Portland cement by CSA cement and a mixture of Portland cement with CSA cement in a ratio of 50% to 50%. The respective creep tests show that the typical creep is greater when using pure Portland cement than when using CSA cement. A mixture of Portland cement and CSA cement shows a further reduction in creep deformation.

All three approaches presented here indicate a lower creep deformation in the presence of CSA or expansive admixtures compared to normal concrete. However, in the literature, there is only insufficient information regarding the effects of creep of the expansive concrete, the decrease in chemical prestressing due to creep, and the general factors influencing the creep processes. Further research should be carried out in this area to be able to make reliable statements regarding the creep behaviour of expansive concrete and to estimate the prestressing force losses resulting from the creep phenomenon.

3.6 Long-term behaviour and durability

Long-term stability

Long-term behaviour and possible prestress losses are an important issue during design of chemically prestressed specimens, especially as the introduced prestress levels are usually not very high. Research conducted by He et al. (2009, 2010) aimed to estimate the long-term behaviour of expansive concrete. Measurements of deformations were conducted on a three-year basis. Concrete prisms with dimensions of 100 x 100 x 550 mm^3 were reinforced with either steel rebars, steel fibres or a combination of both. The concrete mixture was based on sulfoaluminate expansive cement, lime rock coarse aggregate with a size of 5-20mm, medium sand and gravel. The exact amount of expansive agent in cement is not provided, however concrete prestress should be between 3 and 5 N/mm^2). Results proved that losses of prestress introduced with sulfoaluminate expansive cement were neglectable. Indeed, the deformation achieved after about three months was almost the same as the deformation after 36 months (He et al., 2010). The same is confirmed by Nagataki and Gomi (1998), who claim that if the expansive concrete has obtained proper water curing, it does not change the expansive rate after it has reached a settled value.

Freeze-thaw resistance

Freeze-thaw resistance is also an issue that might interfere with practical applications of chemically prestressed elements. Concrete with CSA admixture should have similar freeze-thaw resistance as Portland-cement concrete, unless it has excessive expansion or not enough restraint provided, then the resistance might be lowered (Nagataki and Gomi, 1998).

In research conducted by Wisconsin Department of Transportation in the USA, specimens were placed in 5% NaCl solution after 28 days of curing and then 300 cycles of freezing and thawing were applied. After every 100 cycles loss in mass was measured and compared. After 100 and 200 cycles results for Denka CSA #20 specimen were almost the same as for the control specimen, however, after 300 cycles loss in mass for Denka CSA 20 was equal to 5.4%, whereas for the control specimen it was the only 1.7% (although both still below the 10% threshold permissible in WisDOT Specifications) (Bischoff and Toepel, 2004).

Carbonation of concrete

Also regarding the carbonation of concrete, using CSA admixtures might have a positive effect on the durability. Depth of carbonation measured on plain concrete and expansive concrete samples after 8 and 22 years shown, that depth of carbonation in expansive concrete after 8 years is lower and after 22 years about the same as in normal concrete (Nagataki and Gomi, 1998).

3.7 Mechanism of expansion

Mechanism of expansion of CSA cement is not yet thoroughly understood and thus estimating expansion and prestress is challenging as well (Zdanowicz and Marx, 2019*a*). Because of the gaps in knowledge about the chemical behaviour of CSA during its hydration and multiple factors which influence this behaviour, it is not yet possible to estimate the expansion of a given concrete mixture basing only on its composition. Trial batches and test specimens are necessary before the application of a new mixture or even changes in an existing one. Nevertheless, several hypotheses about the mechanism of expansion are available and are summarized here to show the direction of developments in this area.

3.7.1 Crystal Growth Theory

Crystal growth theory is one of the oldest theories of expansion; it assumes that the growth of the ettringite crystals is responsible for the expansion of calcium sulphoaluminates cement. Various hypotheses were stated regarding the formation of ettringite crystals and how they contribute to expansion (Cohen, 1983). At first, it was believed that ettringite crystals grow on the surfaces of the expansive particles and contribute to exerting expansive force. It assumed that expansive particles (C_3A and C_4A_3S) are spherical and that ettringite ($C_6AS_3H_{32}$) created during hydration process (Eq. 3.2 and 3.3) forms a layer on these particles. With further hydration, the thickness of this ettringite layer increases and pushes other particles apart, thus leading to expansion.

$$C_4A_3\overline{S} + 8C\overline{S} + 6C + 96H \rightarrow 3C_6A\overline{S}_3H_{32} \tag{3.2}$$

$$C_3A + 3C\overline{S} + 32H \rightarrow C_6A\overline{S}_3H_{32} \tag{3.3}$$

This model was later validated and it was found out that ettringite does not form as a layer, but as irregular particles in the first phase and as needle-like crystals later on. The long crystals were believed to later be arranged radially around C_4A_3S particles. Further, it was discovered that although ettringite crystals are indeed formed at the surface of the C_3A particles, they do not remain there but eventually peel off and are later distributed in the solution. Other theories said that not the total ettringite amount was contributing to the expansion and that some of the ettringite was created in existing voids in the concrete structure and did not cause expansion, so there is no direct link between the ettringite amount and expansion rate (Cohen, 1983).

3.7.2 Swelling Theory

Another theory of expansion is based on the assumption that ettringite is formed as colloidal crystals with a gel-like structure. The crystals have a large specific surface area and they absorb water particles on their surface, which leads to exerting swelling pressure and thus to expansion. In the absence of lime, however, colloidal ettringite cannot be formed and the crystals are rather long, which means that their adsorption of water is also lower and does not contribute to expansion (Mehta, 1973). This theory was is also believed not to be valid (Mather, 1984; Bizzozero et al., 2014).

3.7.3 Theory of Crystallization Pressure

Another possibility, a theory of crystallization pressure, checks if a solution is saturated, undersaturated or supersaturated. In the last case, when the logarithm of the ratio of ion activity product to the solubility product is higher than zero, pressure can be exerted by crystallization, and this crystallization pressure finally leads to expansion (Bizzozero et al., 2014). However, crystals growth happens first in the free space (so in the larger pores) without exerting expansive forces, therefore the geometry of the pores and the pore solution also influence the ability of a cement paste to cause crystallization pressure. This influence of pore structure features such as pore size distribution was also recognized in (Chaunsali and Mondal, 2015*a*).

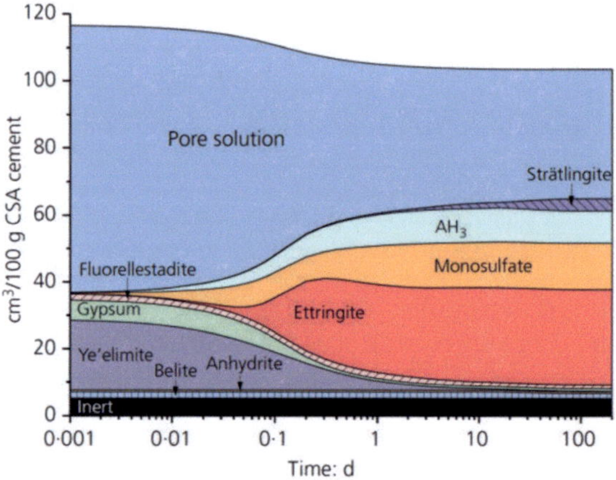

Figure 3.6: Results of thermodynamic modelling of hydration of a CSA cement (Hargis et al., 2018)

Hargis et al. (2018) pursued the path of crystallization pressure as responsible for expansion in CSA cement. They investigated the hydration process and expansion of CSA cements with various calcium sulfate to ye'elimite ratios and raised an important issue – as described above, it is well known, that ettringite contributes to the expansion, but can the whole expansion be attributed only to ettringite? If the theory of crystallization pressure is adopted, any crystallising phase might be contributing to the expansion of CSA concretes. In their study, monosulfate, CAH_{10}, AH_3 and strätlingite were considered and their hydrostatic tensile stress and the potential effect on expansion were calculated, however it is necessary to confirm these effects experimentally to understand why the pressure exerted by some hydrates is higher than by others. Figure 3.6 shows results of thermodynamic modelling of hydration of a CSA cement with a molar ratio of calcium sulfate to ye'elimite of 1 (Hargis et al., 2018). The theory of crystallization pressure remains currently the most viable approach to understand the mechanism of expansion of calcium sulphoaluminate cements or concretes with CSA admixture.

Part II

EXPERIMENTAL RESEARCH

Chapter 4

Experimental programme

4.1 Programme of experimental research

The aim of the work was to find out if the chemical prestressing of specimens with carbon textile reinforcement is possible. For this purpose, it was necessary to conduct numerous material investigations of various concrete mixtures to determine a concrete mixture suitable for chemical prestressing. These investigations included first of all strain measurements (shrinkage and expansion, free and restrained, short- and long-term) to provide a mixture that can produce sufficient stress level (Chapter 6). Secondly, when the choice was already limited to maximum of four mixtures, material tests were conducted: compressive strength of concrete, modulus of elasticity and concrete flexural tensile strength (Chapter 5). Material investigations were followed by structural tests (Chapters 8 and 7) . Firstly, flexural tests were conducted on small-scale slab specimens with carbon textile reinforcement as a preliminary feasibility study. Secondly, large-scale slab specimens with carbon reinforcement were tested in four-point bending tests, followed by analogue tests on beam specimens with steel reinforcement as a validation. Flexural tests of beams and large-scale slabs were carried out up to a deflection of 40 mm to observe the cracking behaviour and behaviour under loading and unloading cycles. Finally, tensile and pull-out tests were conducted on specimens with textile reinforcement as well, to validate the influence of expansive admixtures and chemical prestressing in a plain, uniaxial stress state.

4.2 Materials and specimens nomenclature

4.2.1 Concrete mixtures

The preliminary analysis included concrete mixtures based on CEM I 42.5 N, CEM I 52.5 R, and CEM II/A-LL 52.5 N. Maximum aggregate size ranged between 8 and 16 mm. Two types of superplasticizers were tested, Polycarboxylate ether (PCE) and Poly-Aryl-Ether (PAE), initially also amounts of CSA expansive admixture equal to 6% and 10% of cement weight were tested. After the preliminary analysis, the final mixture was chosen for further experimental part based on the analysis of the expansion process and time of expansion, as well as on the magnitude of expansion and material properties of concrete. It was a self-compacting concrete (SCC) mixture, based on the CEM II/A-LL 52.5 N cement (w/c ratio of 0.38), with an addition of fine limestone to improve its expansive properties (vide Chapter 3.2.1). As aggregates quartz and granite with a maximum size of 8 mm were used. Polycarboxylate ether (PCE) superplasticizer was added to achieve the desired workability. The reference mixture with no expansive admixture is further referred to as C00 mixture, other mixtures have an addition of CSA-based expansive admixture in an amount of either 15%, 20% or 25% based on the cement weight. The detailed composition of the mixture is a subject of privacy agreement with the company involved in the project (Zdanowicz and Hansen, 2019). The slump flow of all SCC mixtures was between 750 and 850 mm and thus the mixtures were classified as consistency class SF3 acc. to EN 206 (EN 206:2013, 2013). Mixing time and regime was constant for all mixtures, as prolonging the mixing time can lead to a decrease in expansion rate (see Chapter 3.2.2). For all mixtures, the total mixing time equalled 360 seconds. The setting time was checked with a penetrometer according to standard ASTM C 403 (ASTM, 2016) to determine the time to start expansion measurements.

4.2.2 Textile reinforcement

In all specimens reinforced with carbon textile reinforcement the same type of textile grid is used, namely *solidian GRID Q142/142-CCE-38* (Figure 4.1) (Solidian GmbH, 2017).

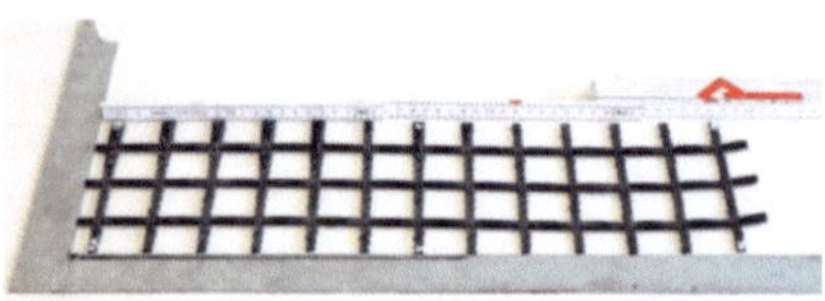

Figure 4.1: Textile reinforcement *solidian GRID Q142/142-CCE-38*

It is a textile material consisting of carbon fibres impregnated with epoxy resin. Its characteristic tensile strength equals 1900 MPa, average tensile strength 2500 MPa and the modulus of elasticity is at least 180 GPa. The axial distance between carbon strands in both directions is equal to 38 mm and the cross-sectional area is 142 mm^2/m. The width of single rovings is around 4-5 mm and thickness about 1-2 mm, so that the cross-sectional area of a single roving equals in average 5.42 mm^2. In all specimens, one layer of the textile reinforcement is located in the middle of their depth.

4.3 Specimens

4.3.1 Nomenclature

In the thesis a unified nomenclature of specimens is used. The initial part of the name tells which concrete mixture was used, that is if the element is a reference, non-prestressed specimen (C00) or a specimen made of expansive concrete (C15, C20, C25). Second capital letter designates the type of test conducted (S and F - flexural tests of large and mid-scale slabs, B - flexural tests of beams, P - pull-out and T - tensile test), and the following small letter *a* - *d* is a designation of a particular specimen (there were always at least two specimens casted for each test series). Specimens are summarized in the Table 4.1, however the table does not include many further specimens used for material tests, tests of concrete properties and expansion measurements which are mentioned in Chapter 5 or in further literature (Zdanowicz and Hansen, 2019; Zdanowicz and Marx, 2019*a*; Zdanowicz et al., 2020).

4.3.2 Geometry

Pull-out specimens (P-specimens)

For pull-out tests, specimens were cast from C00 and C20 mixtures. Their dimensions equalled 450 x 100 x 20 mm^3 and the carbon textile reinforcement was located in the middle of their depth. There were in total three longitudinal carbon rovings going through the specimen. Specimens were stored for 24 hours in the formwork covered with foil at the temperature of 20°C, then removed from formwork and stored for the next 27 days in an environmental chamber with 20°C and 65% RH. They were placed horizontally on the oiled formwork plates to avoid additional restraint and bending.

After 21 days the specimens were prepared for the tests by performing saw cuts as shown in Figure 4.2. The length of the two cuts on both sides, which aim was to cut through the external rovings, equalled 40 mm and was performed 165 mm from the specimen's edge. The cut through the middle roving

Table 4.1: Specimens: nomenclature and dimensions

Specimen name	Concrete	Dimensions [mm]
Flexure – large slabs		
C00_S_(a,b,c)	C00	
C15_S_(a,b,c)	C15	2000 x 1000 x 30
C20_S_(a,b)	C20	
Flexure – mid-scale slabs		
C00_F_(a,b,c)	C00	
C15_F_(a,b,c)	C15	1300 x 150 x 30
C20_F_(a,b,c)	C20	
C25_F_(a,b,c)	C25	
Flexure – beams		
C00_B_(a,b,c)	C00	
C15_B_(a,b,c)	C15	2000 x 160 x 160
C20_B_(a,b,c)	C20	
Pull-out		
C00_P_(a,b,c)	C00	450 x 100 x 20
C20_P_(a,b,c,d)	C20	
Tension		
C00_T_(a,b,c)	C00	
C15_T_(a,b,c)	C15	400 x 100 x 30
C20_T_(a,b,c)	C20	

was also 40 mm long and performed 90 mm from the edge of the specimens so that the specimen was divided into two sections, with a bond length of 75 mm.

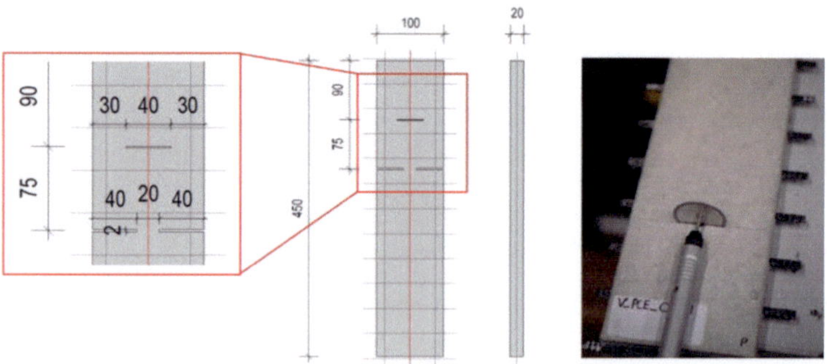

Figure 4.2: Pull-out specimens geometry and performing the middle saw cut

Tensile specimens (T-specimens)

For tensile tests, three specimens were cast out of C00, C15 and C20 mixtures. All specimens had dimensions of 450 x 100 x 30 mm^3 and they were reinforced with a single sheet of a carbon textile reinforcement placed in the middle of their depth (Figures 4.3 and 4.4). Specimens were stored for 24 hours in the formwork covered with foil. After this time, they were removed from the formwork and stored in an environmental chamber with 20°C and 65% RH.

All specimens were provided with DFOS sensors located in the middle of their length (Fig. 4.3). Fibres were introduced to and out of specimens on their sides so that on the middle textile roving the measurement length was equal to ca. 280 mm. Such arrangement ensured that compression caused by clamping of the specimens during tensile tests will not damage the optical fibres.

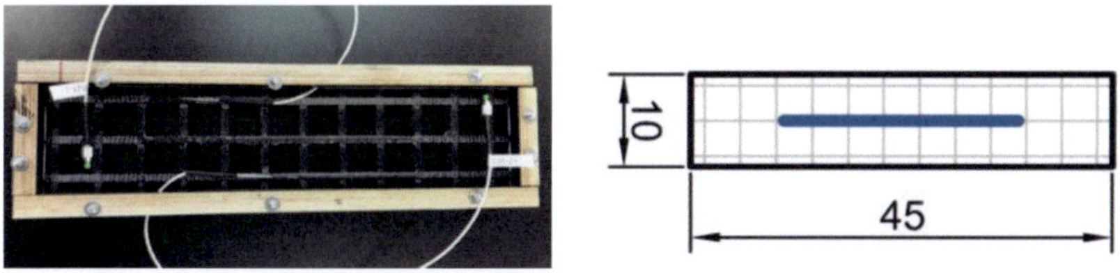

Figure 4.3: Formwork for tensile specimens with installed DFOS sensors and location of DFOS sensors in the specimen

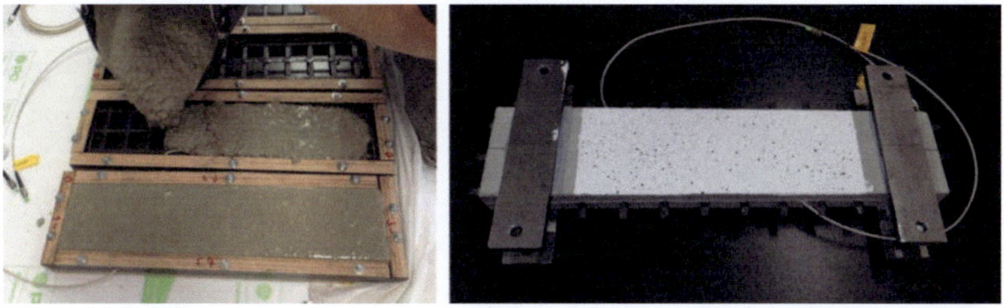

Figure 4.4: Concreting of specimens for tensile tests (left) and preparation for tensile test - clamping steel plates and DIC area visible (right)

Large slabs for flexural tests (S-specimens)

In total, eight slabs were cast, three of concretes C00 and C15 and two of C20 concrete. The dimensions of the slabs were equal to 1000 x 2000 x 30 mm^3

and they were reinforced with a single sheet of a carbon textile reinforcement (*solidian GRID Q142/142-CCE-38* described above), what provided a geometrical reinforcement ratio $\rho_l = 0.47\%$. The dimensions and reinforcement of the slabs were chosen so that they correspond to the current common thickness of textile reinforced facade panels, and so that their reinforcement ratio is equal to the reinforcement ratio of the specimens for restrained expansion test (RET, cf. 6.1). Mechanical reinforcement ratios, that consider geometrical reinforcement ratio and the ratio of tensile reinforcement f_t and compressive concrete f_{cm} strengths (Eq. 4.1), were equal to 0.14 for slabs from C00 series and 0.13 for C15 and C20 series.

$$\omega_l = \frac{f_t \cdot A_t}{f_{cm} \cdot A_c} \tag{4.1}$$

The reinforcement was fixed between two, 15-mm thick timber parts of the sides of formwork to ensure its proper position, additional distance holders were used during concreting in the middle of the slab to avoid leveraging of the reinforcement. Fig. 4.5 present slab formwork with reinforcement and slabs during concreting.

Figure 4.5: Formwork of the slab with installed reinforcement and measuring equipment

Slabs and textile grids were equipped with glass fibres (DFOS Sensors, described in detail in Chapter 6.1.4). In total four sensors were glued to the rovings in each direction, two in the middle cross-section and two in 1/4th of the slab cross-section (named as "G" on the Figure 4.6), while four additional fibres were fixed only on the beginning and end (free sensors, named as "F"), and on their length distanced from the reinforcement with soft foam (to avoid breaking of the sensors). The free sensors should provide measurements of concrete strains. The total length of DFOS sensors equals 15.40 m in the longitudinal direction and 7.20 m in the transversal direction. Additionally, temperature sensors (thermocouples type K) were fixed to measure the temperature during hardening in the middle of the specimens.

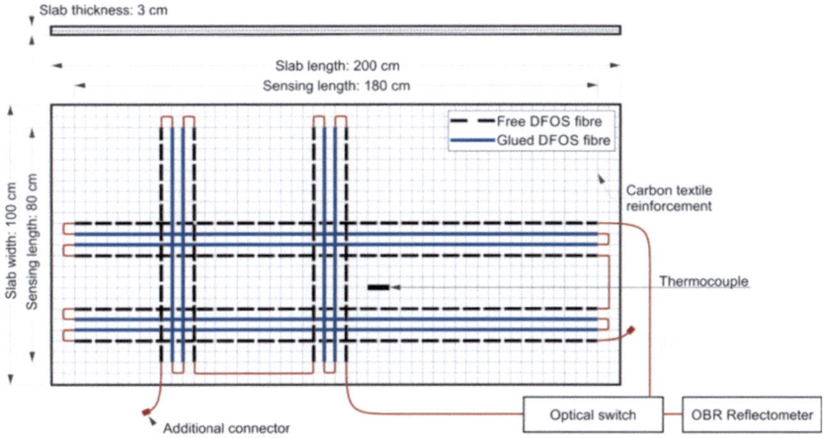

Figure 4.6: DFOS sensors in slabs

Slabs were concreted in horizontal position and stored in the formwork for 48 hours, covered with foil and regularly sprayed with water underneath. After this time specimens were demoulded and stored vertically (Fig. 4.7) in environmental conditions with a constant temperature of 20°C. Prolonged curing in water or moisture-curing was not realized, as it is difficult to realize in industrial applications.

Figure 4.7: Transport and storage of the slabs

Mid-scale slabs for flexural tests (F-specimens)

Mid-scale slabs for flexural tests were cast initially as a feasibility study of chemical prestressing. In total, twelve slabs were made of four concrete mixtures: C00, C15, C20 and C25. Their dimensions were 1300 x 150 x 30 mm^3 and they were reinforced with a single sheet of a carbon textile reinforcement (*solidian GRID Q142/142-CCE-38*), what provided a geometrical reinforcement ratio $\rho_l = 0.47\%$ (the same as in large-scale slabs). Because the textile reinforcement was fixed in the sides of the formwork, no additional distance elements were needed (Fig. 4.8). Specimens were stored under foil for 24 hours and then demoulded and stored in a horizontal position in an environmental chamber with stable conditions of 20°C and 65% relative humidity.

Figure 4.8: Mid-scale slab specimens during concreting

Beams for flexural tests (B-specimens)

Nine beam specimens were cast for flexural tests, three for each concrete type: C00, C15 and C20. The dimensions of the beams were 160 x 160 x 2000 mm^3. All of them were reinforced with single rebar with a diameter of 12 mm, located in the middle of their cross-section, which provided a geometrical reinforcement ratio of $\rho_l = 0.44\%$. The square cross-section, beam dimensions and reinforcement were chosen so that they correspond to the specimens for the restrained expansion test (RET, Fig. 6.1). Each rebar is equipped with three electrical strain gauges placed in the middle of the beam and in the distance of 300 mm to the left and right side, as well as with a glass fibre glued near the longitudinal rib (DFOS Sensor named L0, described in detail in Chapter 6.1.4). On one rebar from each series also a temperature sensor was located. Additionally, there were five further DFOS sensors placed in the beams: two near the top side (E1 and E2), two near the bottom of the beam (E3 and E4) and one in the middle, 20 mm from the rebar (E5). Sensors E1 - E4 were located at a distance of 20 mm from the side surfaces of the beam. The measuring equipment and their nomenclature are shown in the Figure 4.9.

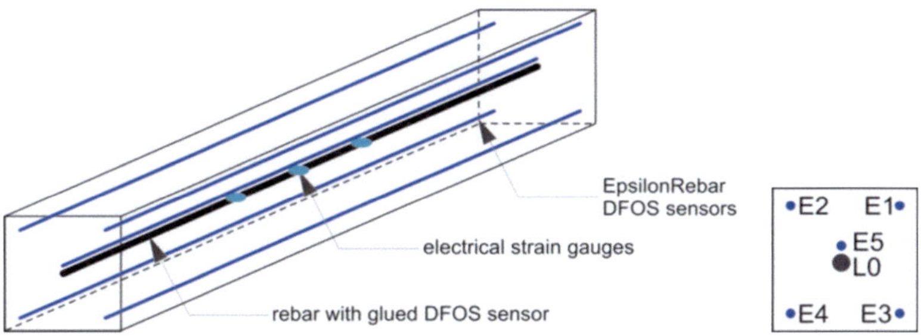

Figure 4.9: DFOS sensors in concrete beams and their nomenclature

Beams were concreted subsequently during 3 days. Figure 4.10 show photos of the sensors installed in the formwork and the beams during concreting. For the first 24 hours, beams were stored in the formwork covered with foil, after this time they were demoulded and placed in environmental conditions with a constant temperature of 20°C.

Figure 4.10: Concreting of beams and the measuring equipment installed in formwork

Chapter 5

Concrete parameters

5.1 Compressive strength

The specimens for testing the compressive strength were prepared in accordance with EN 12390-3 (EN 12390-3, 2019). For this purpose concrete cylinders with a height of 300 mm and diameter of 150 mm were cast and polished so as to have dimensions conforming to the standards. EN 12390-1 specifies a permissible deviation of $0.006 * d$ to ensure levelness, which corresponds to a permissible deviation of 0.09 mm for d = 150 mm (EN 12390-1, 2012). The perpendicularity of the cylindrical specimens was tested in the circumferential direction at 0°, 120° and 240°, whereby the maximum deviation might be 1.05 mm (EN 12390-1, 2012).

After 28 days and appropriate preparation of the samples, the compressive test was carried out according to EN 12390-3 (EN 12390-3, 2019). The specimens were aligned centrically under the testing machine, the loading speed of the strength test under compressive stress was 0.5 N/mm^2 per second. The vertical deformation path was measured with three laser distance sensors in the third points of the circumferential direction at 0°, 120° and 240° and the maximum load was recorded. For compressive strength, at least three cylindrical test specimens were tested for each concrete mix. Figure 5.1 shows a specimen in the test setup and the typical fracture pattern after the test. The results of the compressive strength are summarized in Table 5.1.

In the Figure 5.2 the results of compressive strength are compared for the mixtures with various amounts of the expansive admixture: 0%, 15%, 20% or 25% of the cement by weight (on the x-axis). Specimens marked with "K" were cured in the environmental chamber under 20°C and 65% RH, these with

Table 5.1: Compressive strength of concrete mixtures

Concrete	Compressive strength				
	Value [MPa]	Mean [MPa]	Std. dev. [MPa]	Mean (batches) [MPa]	COV [%]
C00 (batch A)	86.5 85.3 84.2	85.3	1.2	82.7	2.7%
C00 (batch B)	79.1 81.1 80.3	80.2	1.0		
C00 (batch C)	82.1 84.4 81.9 82.1 82.9	82.7	1.0		
C15 (batch A)	87.4 82.7 86.3	85.5	2.5	88.6	3.6%
C15 (batch B)	90.8 90.0 87.0 90.4 89.3 93.7	90.2	2.2		
C20 (batch A)	95.4 93.1 95.4	94.6	1.3	92.6	2.1%
C20 (batch B)	90.3 90.0 89.6	90.0	0.4		
C20 (batch C)	92.2 92.0 93.7 94.1 93.0 91.1	92.8	1.0		
C25 (batch A)	67.7 69.9 67.8	68.5	1.2	68.5	1.8%

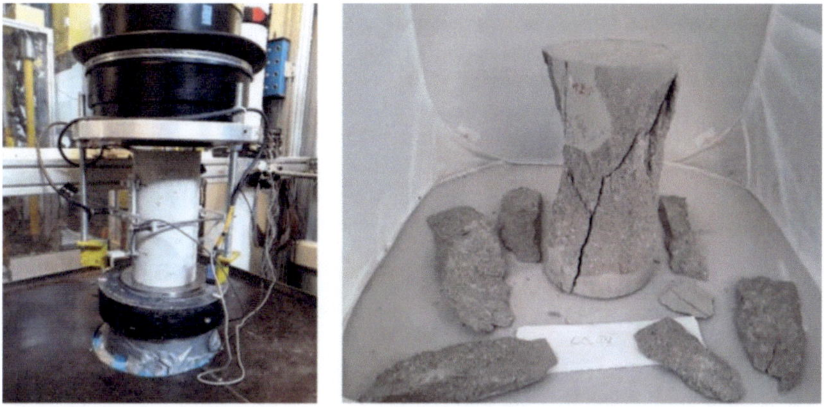

Figure 5.1: Compressive test of a concrete specimen and a picture after failure

"W" for the first 7 days in water at a constant temperature of 20°C.

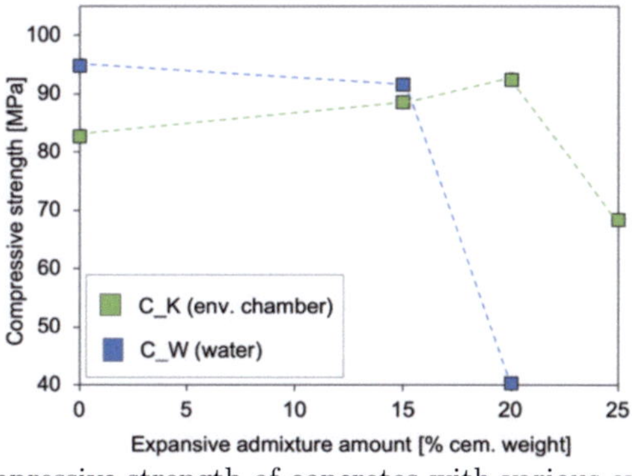

Figure 5.2: Compressive strength of concretes with various expansive admixture amounts

As described in chapter 3.3, there is no uniform opinion of the researchers on the influence of the CSA-based expansive admixtures on the compressive strength of concrete. Basing on 54 compressive strength tests of various concrete mixtures we confirmed the hypothesis that up to a defined threshold (here defined as expansive admixture amount) there is a positive influence of the addition of expansive admixtures on compressive strength (+7.1% or +12% increase for C15 and C20 when compared to reference mixture), however after reaching this threshold a sudden drop in the compressive strength (-17.2% for C25) occurs. Moreover, the threshold is lower for specimens cured under water than for those cured in the environmental chamber - curing in water for a prolonged time leads to a decrease of compressive strength in each mixture with expansive admixture and a severe decrease for specimens where high amounts of CSA are used, what is caused by a deterioration of the internal structure of concrete. Not only microcracks occurred in these cases, but a severe spalling of concrete on its surface was visible.

5.2 Modulus of elasticity

The modulus of elasticity was determined on a further three test specimens. The test of the E-modulus was based on EN 12390-13 (EN 12390-13, 2013), whereby the determination of the secant modulus was carried out with method B. During the test, three load cycles were conducted for each test specimen. The preload stress was equal toσ_{p} = 0.5 N/mm^2 and the upper test stress to σ_{a} = 1/3 of the compressive strength of the previously tested specimens from the same concrete batch. The specimen strains were measured with an extensometer attached to the specimen. The results of the modulus of elasticity are summarized in Table 5.2.

Table 5.2: Modulus of elasticity of concrete mixtures

Concrete	Modulus of elasticity		
	Value [GPa]	Mean value [GPa]	Std. dev. [GPa]
C00	33.35 33.27 33.65	33.42	0.20
C15	32.95 32.19 33.99	33.04	0.90
C20	32.61 32.14 32.11	32.29	0.28

In comparison to compressive strength, for the modulus of elasticity actually no significant influence of the addition of expansive admixture can be observed. The modulus of elasticity of the mixture C15 is 1.1% lower than of C00, and C20 shows a 3.4% drop in E-module.

5.3 Tensile strength

The flexural tensile strength of the C00 and C20 concrete specimens was tested according to EN 12390-5 standard (EN 12390-13, 2013). Prismatic specimens with dimensions of 100 x 100 x 350 mm^3 were concreted for C00 and C20 concrete mixtures and then after 34 days subjected to a three point bending test as shown in Figure 5.3. The force-controlled test was carried out at a loading speed of 0.11 kN/s as specified in the standard. The results of the flexural strength tests are summarized in Table 5.3. The C20 series specimens achieved an average of 6.17 MPa, 18% higher than the C00 series tensile strength (5.21 MPa).

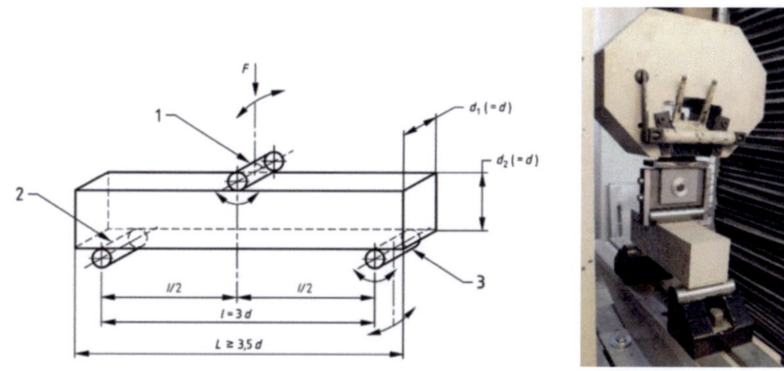

Figure 5.3: Flexural tensile test: test setup (left) (EN 12390-13, 2013) and specimen during test (right)

Table 5.3: Tensile strength of concrete mixtures

Concrete	Max. Load [kN]	Deflection [mm]	Tensile strength Value [MPa]	Mean value [MPa]	Std. dev. [MPa]
C00	16.03	0.80	4.81	5.21	0.36
	18.31	0.97	5.49		
	17.79	0.78	5.34		
C20	20.44	1.55	6.13	6.17	0.27
	21.53	1.73	6.46		
	19.74	0.72	5.92		

5.3.1 Flexural and direct tensile strength

According to Model Code 2010 (fib Fédération internationale du béton, 2013), the relationship between flexural and direct tensile strength of concrete can be calculated with the following formula:

$$f_{ct} = f_{ct,fl} \cdot \frac{0.06 \cdot {h_b}^{0.7}}{1 + 0.06 \cdot {h_b}^{0.7}} \tag{5.1}$$

where:
f_{ct} is the direct tensile strength
$f_{ct,fl}$ is the measured flexural tensile strength
h_b is the height of the element in mm
h_0 is the reference height equal to 100 mm.

As the specimens used in the flexural tensile tests had cross-sectional dimensions of 100 x 100 mm^2, the factor for multiplication is equal to 0.6.

Therefore, for the flexural tensile strength values from the experiments, a direct tensile strength of plain concrete can be calculated:

$$f_{ct.C00} = f_{ct,fl.C00} \cdot 0.6 = 5.21 \cdot 0.6 = 3.13\, N/mm^2 \tag{5.2}$$

$$f_{ct.C20} = f_{ct,fl.C20} \cdot 0.6 = 6.17 \cdot 0.6 = 3.70\, N/mm^2 \tag{5.3}$$

And their ratio is equal to:

$$\alpha_{fct} = \frac{f_{ct.C20}}{f_{ct.C00}} = \frac{3.70\, N/mm^2}{3.13\, N/mm^2} = 1.18 \tag{5.4}$$

Chapter 6

Expansion measurements

6.1 Measurement methods

6.1.1 Restrained Expansion Test

Measurements of restrained expansion, also known as Restrained Expansion Test (RET) can be performed according to the ASTM C878 standard (ASTM C878, 2014). Until now, no European standard is published, however two other guidelines for restrained expansion are also available: in Italy (UNI, 2017) and Japan (JIS, 1999). All of them are based on the same principle.

For RET, concrete samples with dimensions of 80 x 80 x 240 mm^3 are cast in a specially designed steel formwork with a restraining system. This restraining device consists of a centrally placed steel threaded rod with a diameter of ca. 4 mm (reinforcement ratio of 0.49%) and steel end plates. The plates should be placed at a distance of 240 mm (corresponding to the initial length of the specimen) and fixed with screws. Therefore a full bond is simulated. After hardening, concrete specimens including end steel plates are removed from the moulds and measurements of the length of the steel rod are taken regularly using a dial gauge and a comparator bar stored in the same environmental conditions (to eliminate the influence of their changes). From the changes in the length of the steel bar and the resulting elongations, steel rod strains can be calculated according to the following equation:

$$\varepsilon = \frac{l_t - l_0}{l_{tot}} \tag{6.1}$$

where:
l_t is the corrected comparator reading at time t,
l_0 is the corrected initial comparator reading,
l_{tot} is the gage length,
corrected reading means the measured specimen value minus the comparator bar value.

Figure 6.1: Restrained Expansion Test: concreting the specimens and measurements of the comparator bar and of the specimen

6.1.2 Surface strain measurements

Another method for measuring shrinkage and expansion applied in this work is based on measuring the deformation of a plain concrete specimen on its surface, according to (Bunke, 1991). Here concrete cylinders with a diameter of 150 mm and a height of 300 mm were cast and after removing from the moulds provided with cylindrical steel measuring marks with a hole, which were glued to the surface with a distance of 250 mm. The marks were installed in three measuring lines, located every 120° on the circumference of the cylinders (Fig. 6.2). Measurements were taken with DEMEC (demountable mechanical strain gauge) with a measurement range equal to 13 mm and an accuracy of 5 µε. The initial measurement was taken approximately 24 hours after casting and the average value of the three measurements pro cylinder was calculated for each specimen.

6.1.3 Vibrating wire strain sensors

Vibrating wire (VW) strain sensors are a measurement method that is based on the principle that the eigenfrequency of a wire inside the sensor changes with the changes of the length of the wire. Here VW strain sensors (model 4202 from Geokon Company) with a length of 57 mm were used and concreted

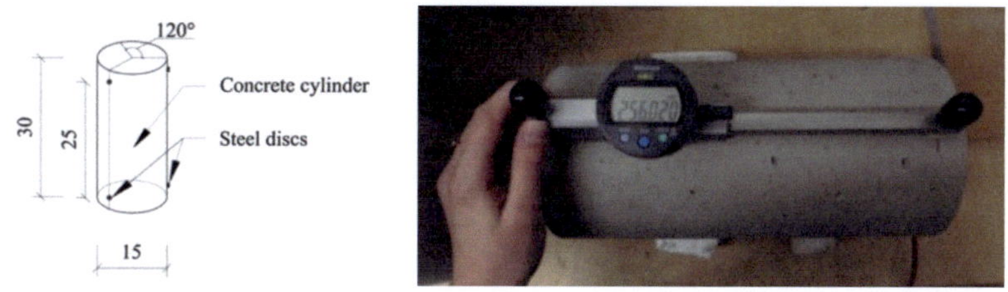

Figure 6.2: DEMEC strain measurements: schema and specimen during measurements

inside cylindrical specimens. The frequency range of the sensor is within 1400–3500 Hz, which allows for strain measurements in the range of around 3000 με with a resolution of 1 με and a system accuracy of 60 με. Each sensor was fixed vertically inside a cylinder, in the middle of its axis (Fig. 6.3) to allow for measurements of the concrete strains inside a specimen. Data were collected automatically every hour for one month and with lower frequency later on. As each sensor is equipped with an integrated thermistor, changes of temperature in the specimens were simultaneously collected.

Figure 6.3: Measurements with vibrating wire strain sensors

6.1.4 Distributed Fibre Optic Sensors

Distributed Fibre Optic Sensors (DFOS) are a novel measurement method that allows measuring length changes and local strains along the previously installed fibres with high precision. The optical fibre technology used previously mostly in the telecommunication field is now finding rising interest within civil engineering branch (Schmidt-Thrö et al., 2016; Sieńko et al., 2018; Fischer et al., 2019) and especially for textile reinforcement (Saidi and Gabor, 2020). The technology is based on light scattering, in this case, Rayleigh scattering, of a signal sent within a specific glass fibre. The glass core has a diameter of 9 µm and is protected with an additional glass cladding with a diameter of 125 µm. The name "distributed" stays in contradiction to conventionally known local measurements, which are available with other methods described above. The technology behind DFOS sensors allows to carry out measurements with a spatial resolution of 5 mm, which for civil engineering purposes can be called continuous in a geometrical sense (Sieńko et al., 2018). Furthermore, high measurement lengths are possible with DFOS, which opens new possibilities for monitoring concrete structures. The post-processing principle which allows quasi-continuous measurements is shown in Figure 6.4: it is possible to choose the sensing range (i.e. sensor length) as well as gage length and sensor spacing, which if chosen as equal values can provide quasi-continuous measurements.

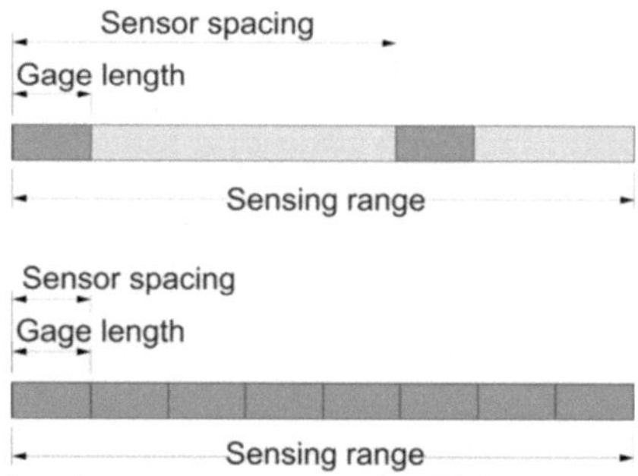

Figure 6.4: Principle of post processing of DFOS data - definition of sensing range, gage length and sensor spacing

Optical sensors were used to monitor strains during shrinkage and expansion in the specimens. Successful application of other types of optical sensors (mainly Fibre Bragg Grating) for shrinkage monitoring is known (Slowik et al., 2004; Habel and Bismarck, 2000; Mufti, 2002; Yazdizadeh et al., 2017), now distributed optical sensors are being applied for such purpose (Sienko et al., 2019; Bassil et al., 2020).

In the thesis, DFOS sensors were installed in concrete in two ways. Firstly, in concrete beams, the *EpsilonRebars* from the SHM Company were used (Fig. 4.9, sensors E1-E5). They have a diameter of 2 mm and a relatively low modulus of elasticity of 3 GPa, so that they should not be working as an additional reinforcement in the specimens, are nevertheless robust enough to ensure easy handling. Furthermore, because of ribs similar to the conventional rebars, the bond between sensors and concrete is provided and the good transfer of strains ensured. Secondly, a DFOS in form of plain glass fibre with a cladding and a total diameter of 0.9 mm (standardised single-mode telecommunication optical fibre SM 9/125) was also used to measure concrete strains in slabs, these sensors were however relatively sensitive to mechanical damage during concreting. The distribution of strains along the fibre was measured with OBR 4600 optical backscatter reflectometer manufactured by Luna Innovations (Luna Innovations Inc., 2013).

DFOS were installed not only in concrete but also on the reinforcement - in beams on steel rebar (sensor L0, Fig. 4.9) and in slabs on carbon textile reinforcement. For this purpose, the surface of the reinforcement was cleaned and a two-component epoxy resin glue was used to fix the fibres. On the steel rebars, sensors were glued near the longitudinal rib. On the carbon textile reinforcement, fibres were glued on the surface of the roving (textile reinforcement impregnated with epoxy resin was used to ensure the compatibility of materials), and the excessive amount of epoxy was removed not to influence the bonding properties. The bending radius of the fibres on the textile grid was approx. 5 - 10 mm, however it was proven before that such a small radius does not influence the transfer of the light signal. Each glued fibre was investigated for losses along its length after application.

In each slab, two fibres were installed - one in the longitudinal direction, which allowed to measure concrete and reinforcement strains in the middle of the slab and 1/4 of its width (see Figure 4.6, in total 15.4 m in each slab), and second in the transversal direction, also in the middle and 1/4 of the slab width (7.2 m).

6.2 Results of free expansion measurements

At the beginning of the investigations, free expansion and shrinkage strains were measured on non-restrained specimens with DEMEC and vibrating wire strain sensors as described above. For this purpose, concrete cylinders were cast from various batches and mixtures. In total, about 100 specimens were cast to assess which mixtures are suitable for further analysis and have the potential of chemical prestressing. Measurements with these methods shown, that results obtained with DEMEC (marked with points) and VW sensors (lines) present a very good agreement (Figures 6.5 and 6.6) (Zdanowicz and Marx, 2018*a*). As the former measurements were conducted on the surface and the latter in the inside concrete cylinders, the small differences between achieved values show also that strain distribution in unrestrained specimens with a depth of 150 mm can be assumed as uniform. Furthermore, it was observed that concrete with 25% of expansive admixture microcracks appeared on its surface, thus it was excluded from further experiments. On the other hand, unrestrained concrete with 15% of EA provided shrinkage compensation, but no chemical prestress unless it was placed in water for the first seven days of maturing.

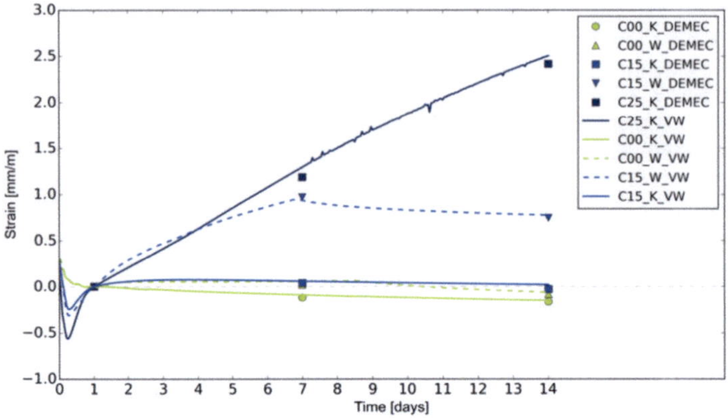

Figure 6.5: Shrinkage and expansion measurements with DEMEC and VW sensors for concretes C00, C15 and C25 during 14 days

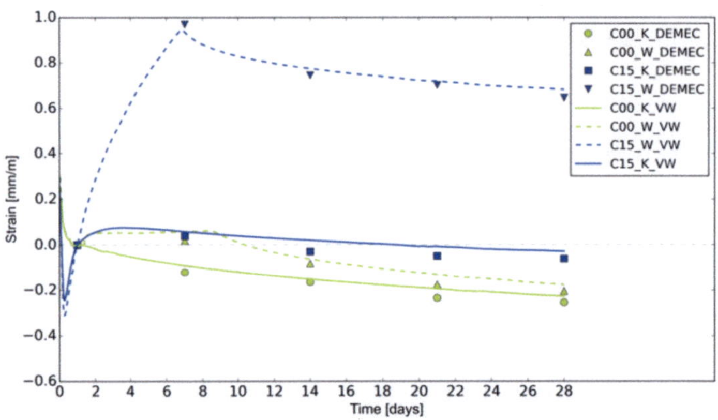

Figure 6.6: Shrinkage and expansion measurements with DEMEC and VW sensors for concretes C00 and C15 during 28 days

6.3 Results of RET measurements

Restrained expansion tests were conducted according to ASTM C878 standard (ASTM C878, 2014). Firstly, four concrete mixtures which were analysed at the beginning of the work were tested, i.e. two mixtures with PCE superplasticizer and two mixtures with PAE superplasticizer. Each of them was made with either 15% or 25% of expansive admixture based on cement weight. These measurements were taken for two years and their results can be seen in Figure 6.7. To assess the stability of expansion, the results after 28 days and after 650 days were compared and are provided in Figure 6.8. Noteworthy, the mixtures with 15% of expansive admixture start to shrink, while 25% mixtures still expand and achieve values that are too high and negatively influence the strength of concrete. In conclusion, PCE superplasticizer was chosen as it exhibited better strain behaviour than specimens with PAE superplasticizer, and it was observed that a huge disproportion is between specimens with 15% and 25% of expansive admixture (notice the logarithmic scale of the Figure 6.8). Therefore the amount of 25% was substituted by an amount of 20% in the next research series.

RET results for the concretes finally used in the experimental programme for this thesis are presented in Figure 6.9. Measurements were taken on three specimens prepared for each series and the mean values are presented. C00 concrete specimens achieve an average value of 0.308 mm/m, C15 of 0.139 mm/m and C00 of -0.401 mm/m.

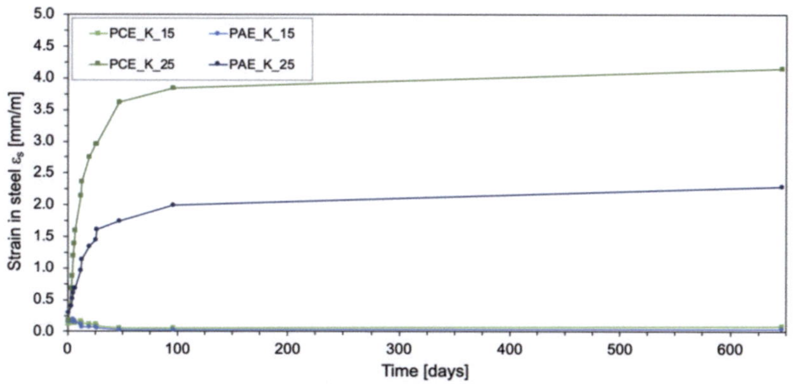

Figure 6.7: RET measurements for concretes C15 and C25 during two years

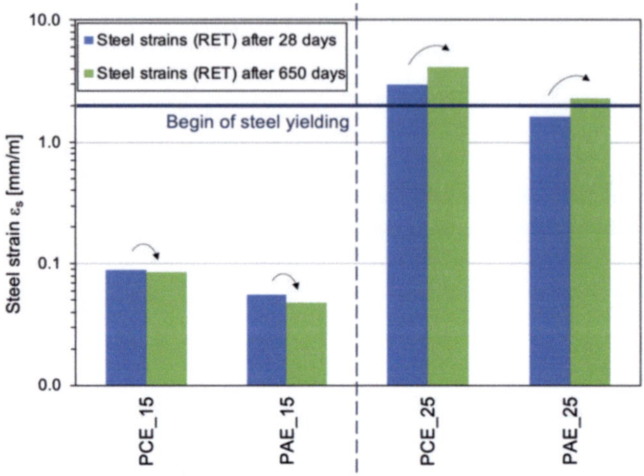

Figure 6.8: Comparison of RET measurements for concretes C15 and C25 after 28 days and two years. Note: the Y-axis is in log-scale

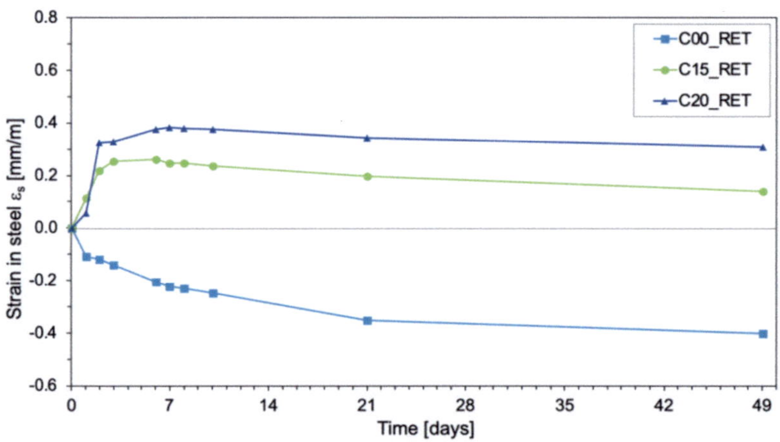

Figure 6.9: RET measurements for concretes C00, C15 and C20

6.4 Results of measurements with DFOS

6.4.1 Expansion of beams

Figure 6.10 shows concrete strain development within 24 days for points located in the middle of the E5 sensors in each beam. There is a good agreement between beams within each series. Beams made of C20 concrete achieve a maximum value of 0.779 mm/m and end values of 0.706 mm/m or 0.716 mm/m, while beams made of C15 concrete achieve expansion strains with a maximum of 0.510 mm/m and of C00 concrete shrinkage in a magnitude of -0.418 mm/m. Figure 6.11 presents additionally the temperature development within 48 hours of the beam specimens. The temperature of the beams made of expansive concrete is slightly higher than that of reference concrete and the peak value is achieved quicker.

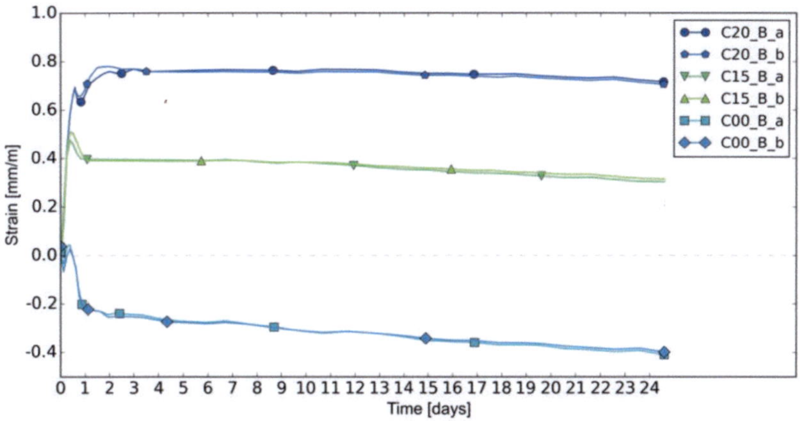

Figure 6.10: Strain development in time of the middle point at sensors E5 in beam specimens

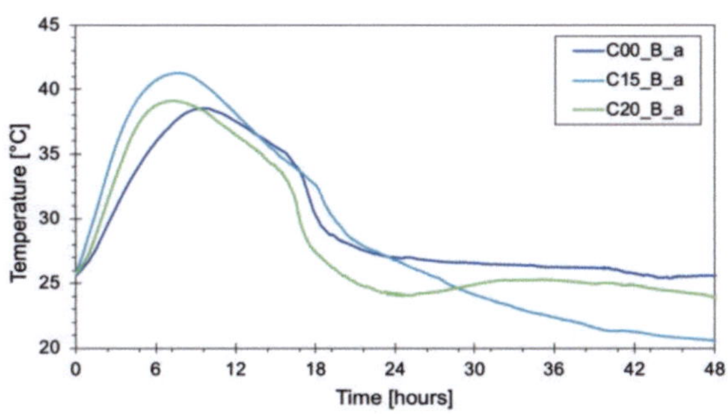

Figure 6.11: Temperature development within 48 hours after casting of the beam specimens

Figures 6.13, 6.14 and 6.15 show the distribution of strains (y-axis) along the length of the beams (x-axis) and strains development in time (within 240 hours, subsequent measurements distinguished with intensifying colours). Sensors in each beam are named according to Figure 4.9.

For C00_B_a beam it is noticeable, that higher shrinkage occurs near the top surface of the beam (sensors E1 and E2) than near the bottom surface (sensors E3 and E4). It is because beams were stored on the ground without any distance elements and therefore the bottom surface was better protected from moisture transfer and evaporation than the top surface. Also, single rebar cannot provide a good restraint for the whole concrete cross-section. Thus, the same can be observed also for beams with expansive admixture. Figure 6.12 shows the strains development during 24 days for all sensors measuring concrete deformations in chosen beams from all series. Concrete further from the rebar tend to deform more (strains in both top sensors are higher) while deformation of the bottom surface is limited by self-weight and being placed on a rough surface (so the strains in both bottom sensors are lower than in the middle of the beam).

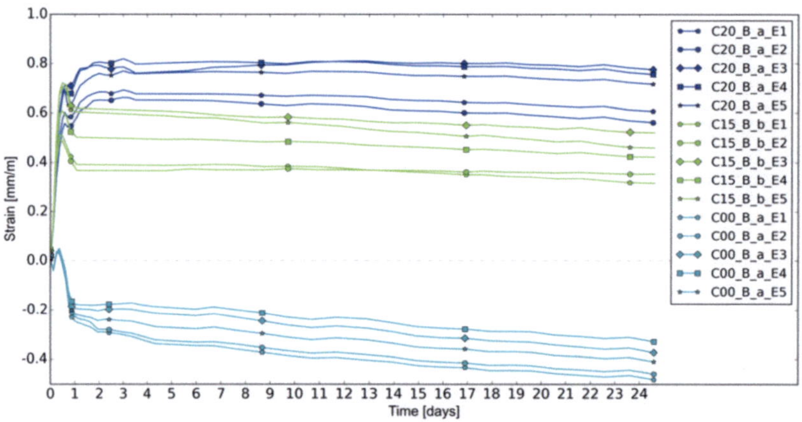

Figure 6.12: Strain development in time of the middle points at all sensors in beam specimens, E1 and E2 near the top surface, E3 and E4 near the bottom surface and E5 in the middle

For all beams we observe also some local differences in strain values and disturbances along the measurement lengths. It might be caused by a chosen post-processing parameters: for the analysis, a 10-mm raster was chosen (Fig. 6.4), what if compared with the maximum aggregate size of 8 mm is a rather low value. Local concentrations of higher or lower strain values are thus attributed to the presence of aggregate in these specific points and could possibly be avoided by choosing a larger gage length and sensor spacing.

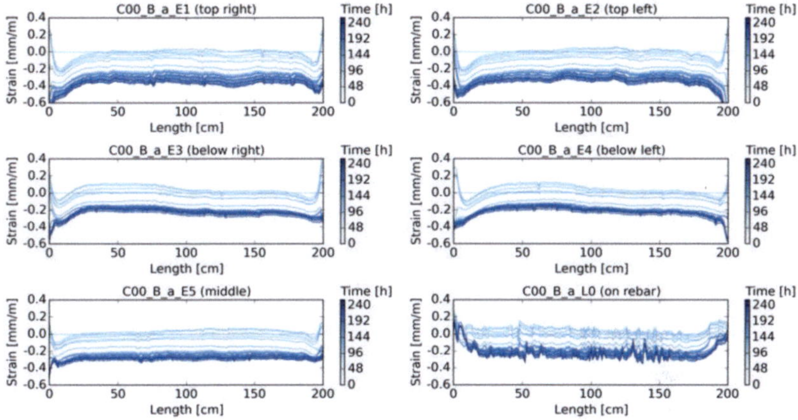

Figure 6.13: Strains development along all sensors in beam C00_B_a within 240 hours

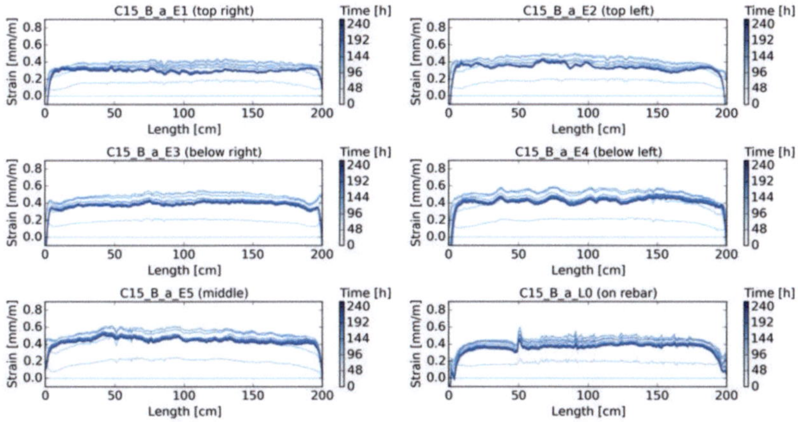

Figure 6.14: Strains development along all sensors in beam C15_B_a within 240 hours

Another useful observation can be made if strains measured with *Epsilon-Rebar* sensor located in the middle of the beam (E5) are compared with strains of the steel rebar (L0). These two sensors are located at 20 mm distance from each other. In direct comparison, a distance after which concrete

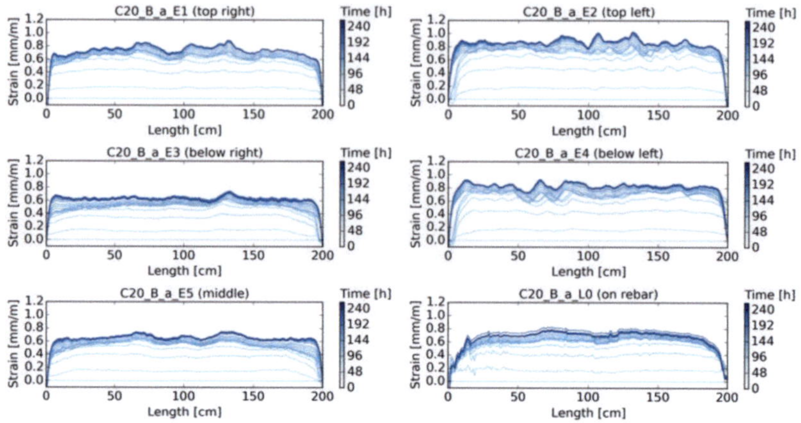

Figure 6.15: Strains development along all sensors in beam C20_B_a within 240 hours

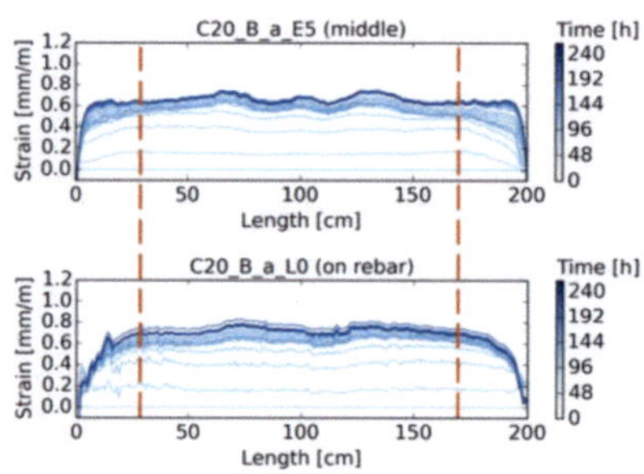

Figure 6.16: Development length for C20_B_a beam - transfer of strains from concrete to the reinforcement

strains are completely transferred to the steel reinforcing bar, so-called development/transition length, can be read from the graphs (visible e.g. in Fig. 6.16 for C20_B_a beam). It is a vital finding for chemically prestressed specimens, as it makes it possible to very precisely define the length after which the full designed prestress is achieved. Such accurate determination of transfer length was difficult to obtain with conventional techniques and offers now valuable insight. For all the beams, this development length is lower than 25-30 cm (Zdanowicz and Marx, 2019*b*). It is an important observation as it differs from what was known until now, that "in chemically prestressed members, expansive strain can be treated as almost uniform in the section more than 500 mm from the member end" (Hosoda and Kishi, 2001).

6.4.2 Expansion of tensile specimens

Similarly, DFOS sensors were located in the specimens prepared for tensile tests, as shown in Figure 4.3. They were located in the middle 250 mm out of 400 mm total length, to avoid damaging sensors during screwing together the steel plates for tensile tests setup. Therefore, the measurement length on Figures 6.17 - 6.18 is equal to 250 mm. In these figures, we can observe the development of strains along the measurement length during the first 144-168 hours (due to technical problems measurements of C00 specimens were started only after removing specimens from the formwork). It is visible that strains are rather constant along the measurement length, however in some of the specimens, especially C00_T_b, small disturbances might be observed in the locations of transversal rovings of the textile grid.

The comparison of all tensile specimens shown in the figure 6.19 proves that specimens made of C00 concrete exhibit shrinkage up to a maximum value of -0.193 mm/m (although the real value should be higher, as strains were not measured during the first 24 hours), while C15 and C20 specimens expand with end values after fours days equal to 0.167 mm/m and 0.444 mm/m, respectively. These values correlate well with the results of RET measurements provided in Figure 6.9 but they are lower than values measured in beams, which can be caused by a very short development length of these specimens. The values are however necessary to later assess the influence of chemical prestress introduced on tensile force values achieved in static tests.

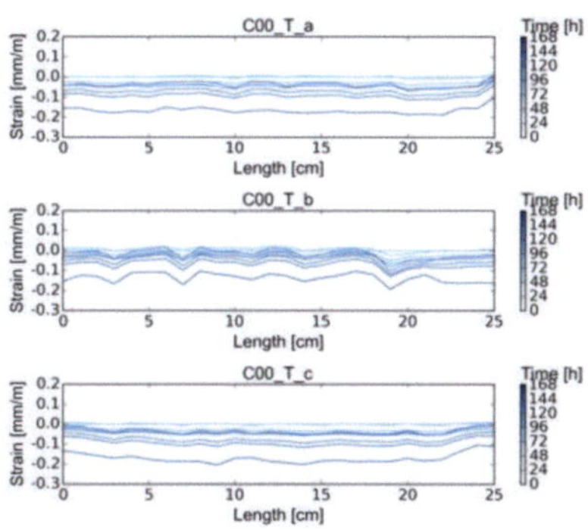

Figure 6.17: Strains development along sensors in tensile specimens C00_T within 144 hours

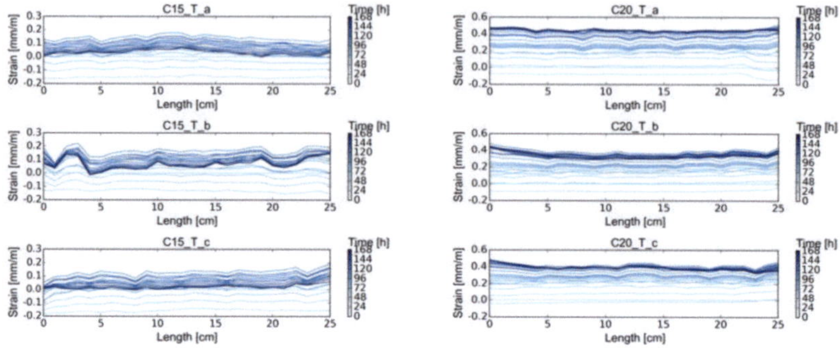

Figure 6.18: Strains development along sensors in tensile specimens C15_T and C20_T within 168 hours

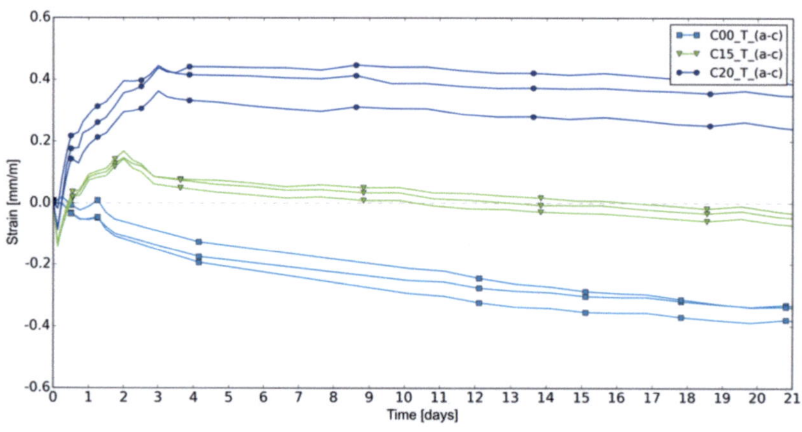

Figure 6.19: Strains development in the middle point of all tensile specimens within 21 days (C20 results will be prolonged)

6.4.3 Expansion of slabs with textile reinforcement

Fibre optic sensors measured strains due to shrinkage and expansion also in large slabs in each direction (longitudinal and transversal). Figures 6.20 and 6.21 show the strains development in time of the specimen C00_P_a in longitudinal and transversal direction within 72 hours after casting. Each line represents a middle point of one sensor, therefore we see results from four glued and four free sensors. Strains measured with sensors that were glued to the carbon rovings are in both cases higher, what suggests that reinforcement could to some extent withstand shrinkage. The next figure, Fig. 6.22 presents also strain development in one of the specimens from series C15, namely C15_P_a, in the transversal direction. Also here we can see the difference between glued and free sensors - free sensors measuring concrete strains show higher values than glued sensors, what implies that reinforcement was acting as a kind of restraint for expansive concrete. Presenting strain development in time along free and glued transversal sensors of this specimen (Fig. 6.23) allowed to investigate the difference between strains in concrete and of the reinforcement. While strain course in concrete was rather smooth, on the texitle grid multiple peaks were observed, and the peaks were located on the crossings of longitudinal and transversal reinforcement. Therefore, an influence of the reinforcement in perpendicular direction on the chemical prestressing can be assumed and confirmed.

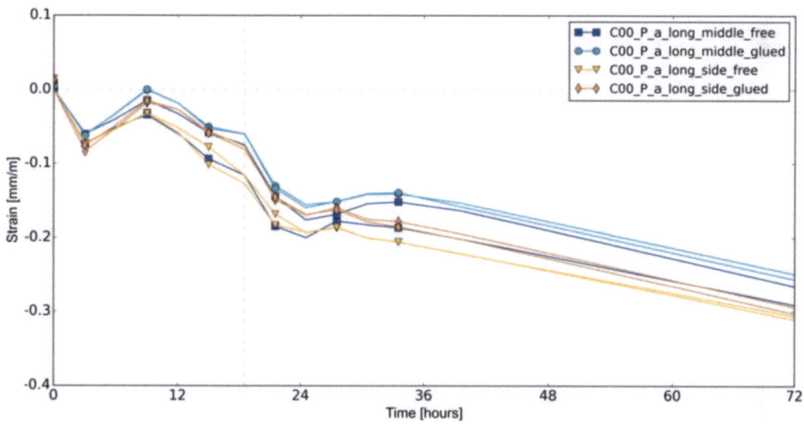

Figure 6.20: Strains development in the middle point of all longitudinal sensors of specimen C00_P_a

Finally, those slab specimens are compared with each other in Figure 6.24 where strain measurements within 24 days are presented. Specimens made of C00 concrete exhibit shrinkage and its value after 24 days is equal to around -0.55 mm/m. For C15 specimens, the mean maximum values achieved during the expansion period are 0.41 mm/m for the transversal and 0.38 mm/m for longitudinal direction, while end values are approximately -0.08 mm/m for both directions. The results from C20 specimens should not be compared here as there was a technical problem with the measurements.

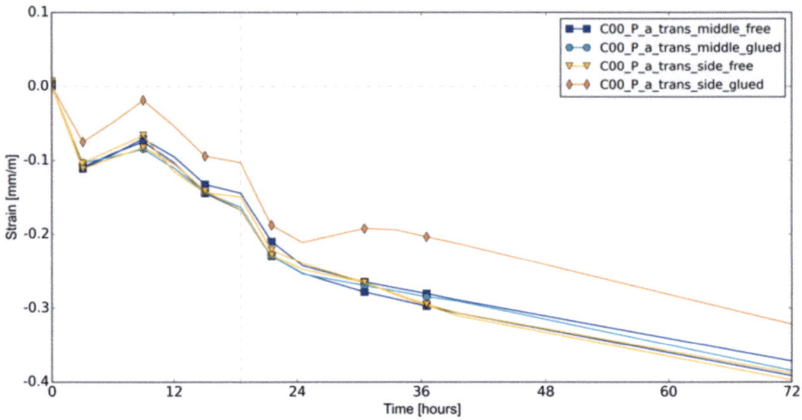

Figure 6.21: Strains development in the middle point of all transversal sensors of specimen C00_P_a

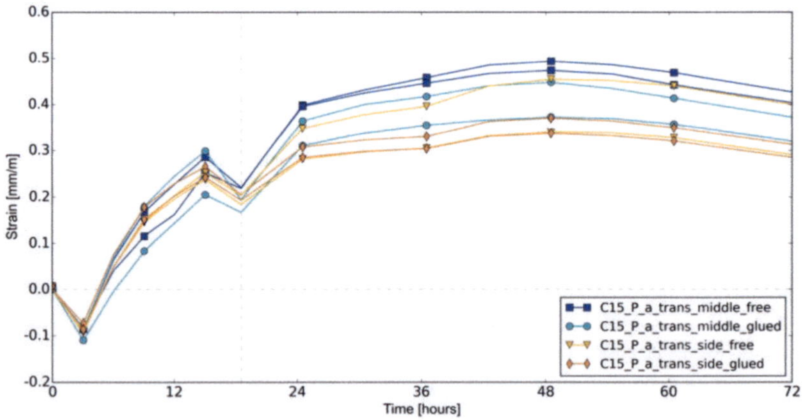

Figure 6.22: Strains development in the middle point of all transversal sensors of specimen C15_P_a

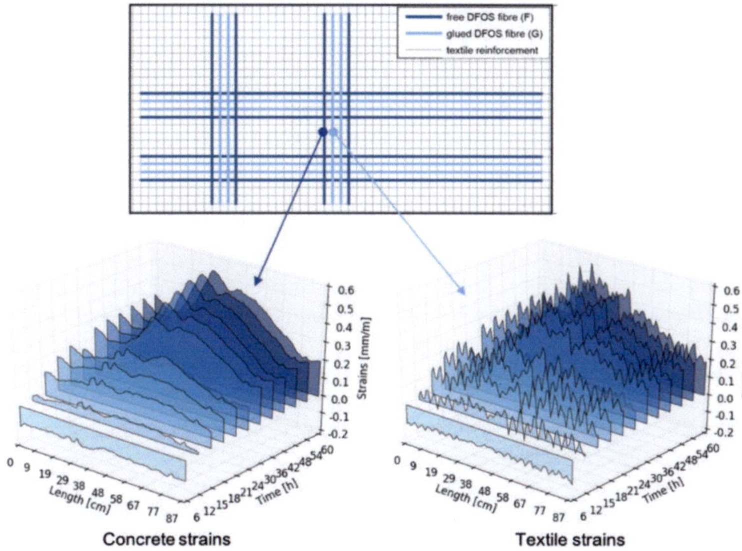

Figure 6.23: Strains development along free and glued transversal sensors of specimen C15_P_a

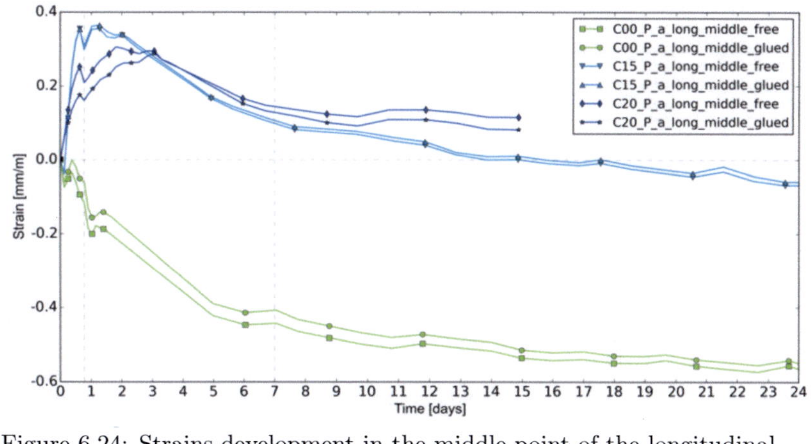

Figure 6.24: Strains development in the middle point of the longitudinal sensors in specimens C00_P_a, C15_P_a and C20_P_a (C20 results break at day 15)

6.5 Summary

For the reference C00 specimens, shrinkage in magnitude between -0.4 and -0.6 mm/m was measured after >20 days. On specimens with textile reinforcement (T- and S-type) it was observed that shrinkage could be compensated with 15% of expansive admixture (C15 mixture). However, for beams and RET specimens, with prismatic shape and steel reinforcement, 15% of EA could introduce expansion strains in magnitude between 0.2 and 0.4 mm/m, which reached maximum value within 7 days and after this time remained nearly constant. Short tensile specimens and slabs made of C15 concrete did not show any expansion after 24 days. Specimens made of C20 concrete exhibited expansive strains, for beams reaching 0.6 - 0.8 mm/m and for tensile elements about 0.4 mm/m. In slabs from C20 series measurements were not correct and the strains could not be estimated.

Furthermore, measurements with DFOS sensors allowed to determine the development length in beams, after which strains of concrete are fully transferred to the steel reinforcement. Noteworthy, this development length reached about 25-30 cm in all series and is much lower than 50 cm assumed until now according to literature (Hosoda and Kishi, 2001).

In slabs with textile reinforcement it was visible that strains in the reinforcement were slightly lower than in concrete, it was true for both shrinkage and expansion strains what implies a good transfer of strains between concrete and the reinforcement.

Expansion strains measured on the reinforcement can be used to assess the prestressing degree of the elements. For example, in beam elements after 24 days the reinforcing steel strains due to expansion reached approximately 0.350 mm/m for the C15 mixtures and 0.750 mm/m for the C20 mixtures, whereas C00 specimens shrinkage about -0.40 mm/m. It means that concrete prestress will achieve about 0.66 N/mm^2 for C15 and 1.01 N/mm^2 for C20 concrete elements.

In tensile elements, prestress can be calculated also basing on strain measurements. After 21 days, C00 specimens reached shrinkage with a magnitude of -0.367 mm/m (mean value), while C20 specimens achieved average estimated expansion of 0.410 mm/m. The difference in strains between C00 and C20 specimens is thus 0.777 mm/m and it corresponds with the concrete prestress equal to 0.75 N/mm^2:

$$\Delta\epsilon_{tex} = 0.777\, mm/m \tag{6.2}$$

$$\Delta\sigma_{tex} = \Delta\epsilon_{tex} \cdot E_t = 0.777\,mm/m \cdot 180000\,N/mm^2 = 140\,N/mm^2 \tag{6.3}$$

$$\Delta F_{tex} = \Delta\sigma_{tex} \cdot A_t = 140\,N/mm^2 \cdot 16.26\,mm^2 = 2.27\,kN = \Delta F_c \tag{6.4}$$

$$\Delta\sigma_c = \frac{\Delta F_c}{A_c} = \frac{2270\,N}{3000\,mm^2} = 0.75\,N/mm^2 \tag{6.5}$$

Chapter 7

Pull-out and tensile tests

7.1 Pull-out tests

7.1.1 Test setup

Pull-out tests were conducted on specimens cast from C00 and C20 mixtures (Zdanowicz, Schmidt, Naraniecki and Marx, 2019). Currently, there is no standardized method for investigating the bond behaviour in textile-reinforced concrete. Hence, the comparative experimental investigation of the bond behaviour between carbon textile and concrete with and without expansive admixture was based on the asymmetrical pull-out tests developed by (Lorenz and Ortlepp, 2011; Lorenz, 2014; Williams Portal, 2015). With the specimen design described in chapter 4.3.2, the bond between the central longitudinal roving and the surrounding concrete matrix over the bond length of 75 mm was specifically analysed. The bond length was chosen in such a way that the roving failure could be excluded. In the longer section of the specimen, sufficient anchoring of the tested roving was ensured over a length of 295 mm.

Two clamping devices were attached to the top and bottom of the specimen for the load introduction into the specimen, see Figure 7.1. The specimen was clamped in the upper and lower area over a length of 30 mm outside of the investigated bond length. On the upper side, a flexible, unrestrained joint was created by a load introduction construction with an integrated load cell to avoid any eccentricities.

The test was carried out in the electromechanical testing machine, whereby the tensile stress was measured with an additional 25 kN load cell at the

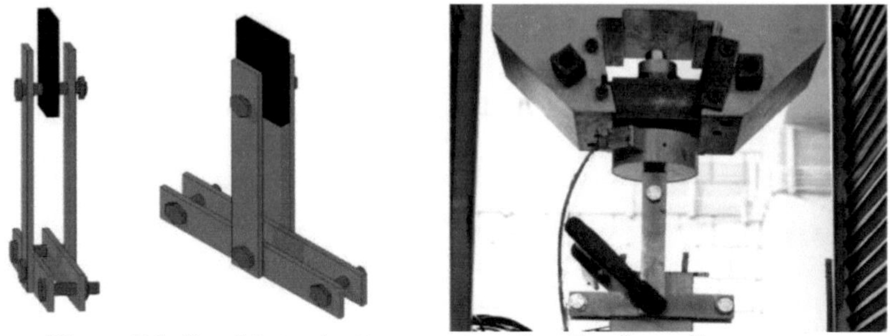

Figure 7.1: Load introduction construction for the pull-out test

upper side. It was possible to ensure a substantially centric load even for a longitudinal roving lying non-centrally in the concrete cross-section by a possible horizontal shift of the upper load introduction plate with flexible joints. Four laser distance sensors with a measuring range of 60 mm and a resolution of 8 µm were used to measure the crack opening. Two lasers were mounted on each side of the specimen above the predetermined breaking point. As reflectors for the lasers, small measuring plates were mounted below the central notch, as cracks were expected to form between the predetermined breaking point and the central notch. The measured values of all laser distance sensors were recorded at 10 Hz. In addition to the crack opening and load, it should be possible to check during the test whether the load introduction structure slips off the specimen surface. For this purpose, the slip was recorded with two additional lasers installed 15 mm above or below the load introduction construction. Figure 7.2 shows the arrangement of the specimen and lasers.

The pull-out tests were carried out with a load speed of 0.5 mm/min until a crack opening $w = 5$ mm was reached. The tested roving was then completely pulled out with a loading speed increased to 10 mm/min. Finally, the length and surface of the roving were visually checked. Besides the load F and the machine path s, the crack opening w in the area of the predetermined breaking point was measured. The measured load was reduced by the dead weight of the upper load introduction and the upper part of the specimen so that the actual pull-out force was then used for further evaluation.

7.1.2 Pull-out force - crack width relationship

The result of the pull-out tests is a pull-out force - crack opening curve for each of the four laser distance sensors. The pull-out force corresponds to the pull-out strength of the examined roving in the area of the shorter bond length in the lower part of the test specimen. The crack opening w is composed of

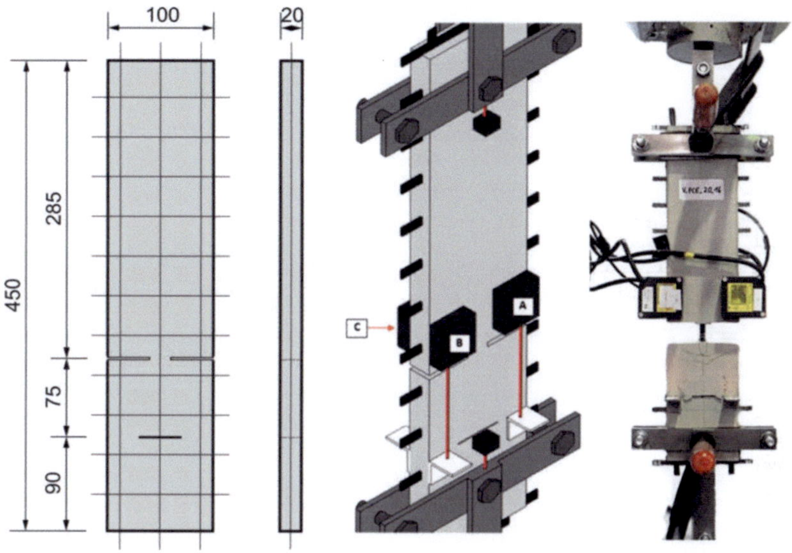

Figure 7.2: Dimensions of the pull-out specimens (left), laser distance sensors arrangement (middle) and specimen during test (right)

the relative displacements of the two specimen parts and the load-dependent deformation of the roving. Figure 7.3 shows an example of a pull-out force - crack opening curve.

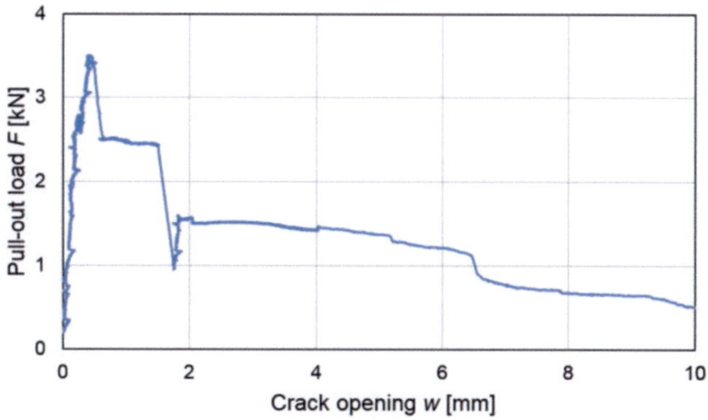

Figure 7.3: Pull-out force - crack opening curve for one of the C00 specimens

All pull-out force-crack opening curves show a fundamentally similar be-

Table 7.1: Characteristic points of the C00_P_(a-c) series in pull-out tests

Specimen	a	b	c	Mean value	Std. dev.
F_H	3.48	4.38	3.26	3.71	0.235
w_H	0.40	0.65	0.35	0.47	0.017
$F_{R1,A}$	2.49	2.58	2.41	2.49	0.005
$w_{R1,A}$	0.65	1.05	0.55	0.75	0.047
$F_{R1,E}$	2.43	2.61	2.36	2.46	0.011
$w_{R1,E}$	1.60	2.75	1.25	1.87	0.411
F_B	1.01	0.42	1.38	0.94	0.156
w_B	1.65	3.10	1.30	2.02	0.607
$F_{R2,A}$	1.54	1.30	1.53	1.46	0.012
$w_{R2,A}$	1.85	3.20	1.40	2.15	0.585
$F_{R2,E}$	1.37	1.30	1.17	1.28	0.007
$w_{R2,E}$	5.00	5.00	5.00	5.00	0.000

haviour: after an almost linear increase, the pull-out force decreases to a relatively constant value over a certain crack opening increase (first plateau). Subsequently, the force that can be absorbed reduces significantly and again remains at a relatively constant level (second plateau) until finally a limit load is reached. All pull-out curves show about the same course, which can be illustrated as shown in Figure 7.4. In this illustration, significant points are indicated whose value pairs of force and crack opening are specific for the investigated specimens. These characteristic points are then compared in the following evaluation for the samples examined (Table 7.1 and 7.2).

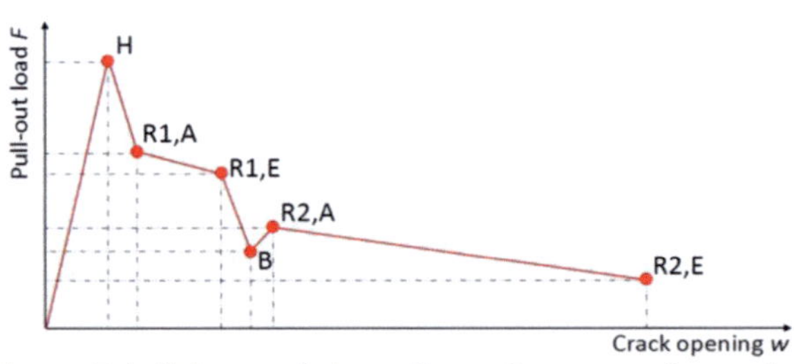

Figure 7.4: Scheme of the pull-out force - crack opening curve

This three-part bond behaviour of the textile reinforcement can be observed in the pull-out force crack-opening curves of all specimens (Figures 7.5 and 7.6). Initially, there is a linear increase in the pull-out force along with a small crack opening, which can be interpreted as an adhesive bond. The time of the first crack cannot be determined precisely from the measured data.

Table 7.2: Characteristic points of the C20_P_(a-d) series in pull-out tests

Specimen	a	b	c	d	Mean value	Std. dev.
F_H	4.98	4.69	4.33	4.47	4.62	0.060
w_H	0.40	0.35	0.40	0.30	0.36	0.002
$F_{R1,A}$	-	3.03	3.34	3.35	3.24	0.022
$w_{R1,A}$	-	0.70	0.60	0.55	0.62	0.004
$F_{R1,E}$	-	2.73	3.02	3.36	3.04	0.066
$w_{R1,E}$	-	1.30	1.00	0.95	1.08	0.024
F_B	1.20	1.77	1.75	2.65	1.84	0.270
w_B	1.15	1.35	1.15	1.00	1.16	0.015
$F_{R2,A}$	2.78	2.08	2.41	2.83	2.53	0.092
$w_{R2,A}$	1.30	1.45	1.25	1.10	1.28	0.016
$F_{R2,E}$	1.37	0.86	1.88	1.90	1.50	0.182
$w_{R2,E}$	5.00	5.00	5.00	5.00	5.00	0.000

However, based on the pull-out force crack-opening curves, it is assumed that this coincides at the latest with the maximum bond load F_H. As a result of the crack formation, the pull-out force is suddenly transferred to the roving that has been cut free and presumably also leads to the adhesive bond being exceeded at the same time. This damage of the interface between the roving and the surrounding material becomes apparent with the falling branch until the first plateau is reached. At this point, the adhesive force decreases due to the decreasing adhesion surface. After the roving has been detached from the surrounding concrete matrix, the pull-out force stays on the first plateau, which can be explained by a still existing friction and shear bond. Therefore, only a small loss of force is observed despite the increasing crack opening. This loss of force is due to the reduction of the bond length as the roving pull-out continues. The measured force decreases after the end of the first plateau before a second plateau is formed after a slight increase. As a possible explanation for this behaviour, the separation of the tested roving from the two existing transversal rovings at the crossing points can be considered. This detachment of the crossing point results from the tearing of the sewing threads running on the transversal rovings and damage to the secondary coating applied in this area. The second plateau can be interpreted, just like the first plateau, as force absorption due to friction bond. Here, however, contrary to the first plateau, only the single tested roving acts with its contact surface without connection to the transversal rovings, resulting in a lower pull-out force. The pull-out force is reduced to zero by the end of the test.

To be able to make a better comparison of the averaged pull-out force-crack-opening curves both are combined in Figure 7.7. Basically, these two averaged curves show the same course. However, the higher pull-out forces are clearly visible in the C20-specimens made of expansive concrete, it can be thus concluded that the bond strength of expansive concrete is also higher.

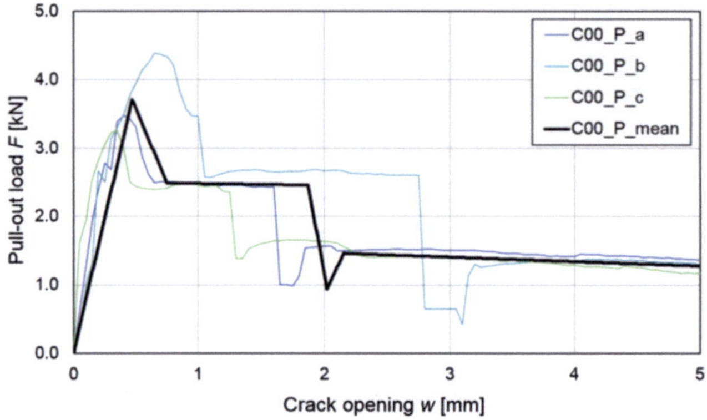

Figure 7.5: Pull-out force - crack opening curves for the specimens from C00 series

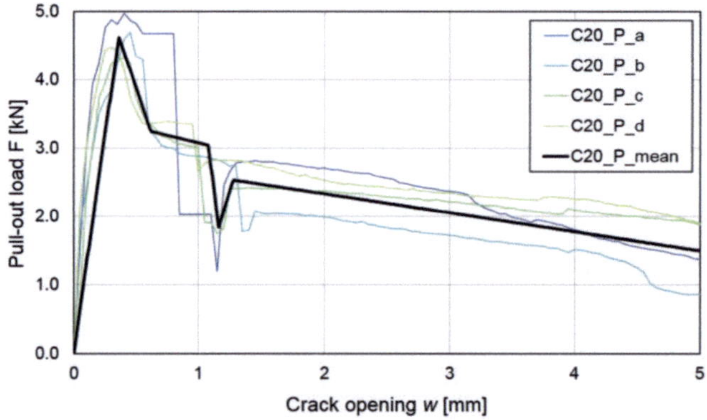

Figure 7.6: Pull-out force - crack opening curves for the specimens from C20 series

7.1.3 Maximum bond strength

As already presumed on the basis of the experimentally determined pull-out force crack-opening curves, the bond strength of the specimens with expansive

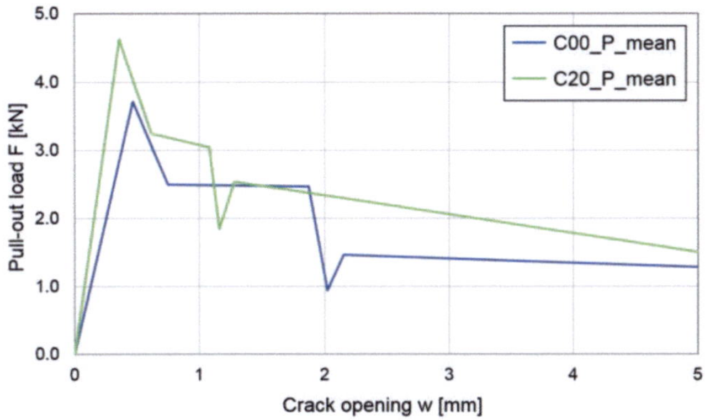

Figure 7.7: Comparison of the pull-out force - crack opening curves for the specimens from C00 and C20 series

concrete shows higher values. The maximum value achieved is approximately 5.99 N/mm^2 for C00 series and 7.46 N/mm^2 for C20 series. Both values reflect the achievement of the maximum bond and were calculated in the following way:

$$\tau_P = \frac{F_H}{A_{bond}} \tag{7.1}$$

The area of a single roving as declared by the producer is equal to 5.42 mm^2, and the cross-section will be here assumed as round instead of elliptic. Thus, the diameter of a single roving can be assumed as $d_{roving} = 2.63$ mm, and the bond area with the bond length equal to L = 75 mm as 619 mm^2. The maximum bond strength calculated for both series is shown in Table 7.3.

Table 7.3: Maximum bond strength

	F_H [kN]	$\tau_{cr.T}$ [N/mm^2]
C00_P_mean	3.71	5.99
C20_P_mean	4.62	7.46

7.2 Tensile tests

7.2.1 Test setup

Tensile tests were conducted on specimens cast from C00, C15 and C20 mixtures. The specimens were equipped with DFOS sensors to measure strain

development in the reinforcement during tensile tests. Furthermore, the Digital Image Correlation (DIC) method was used to capture the crack pattern in the central area of the specimens (between clamping steel parts, on a field with a length of ca. 300 mm) using the setup shown in Fig. 7.8.
Tensile tests were carried in the electromechanical testing machine, where the tensile load and machine displacement were recorded. An additional load cell with a measuring range of 50 kN was installed on the upper side. A flexible, unrestrained joint was installed on the upper side to avoid eccentricities and provide a uniaxial stress state in the specimens. The test setup was oriented at the RILEM recommendations and the dimensions adjusted to the needs of another geometry and reinforcement of specimens (RILEM TC 232-TDT, 2016). The tensile tests were carried out as static tests with four loading and unloading cycles with the following procedure: first cycle in a linearly-elastic area up to a load of 5 kN, second cycle up to 10 kN and then up to maximum load at first crack. After each loading part, the unloading took place up to a load level of 2 kN with the same speed of 0.25 mm/min. The third unloading took place after three or four cracks appeared and a final cycle was to observe the tensile failure of the specimen and pulling out of the reinforcement.

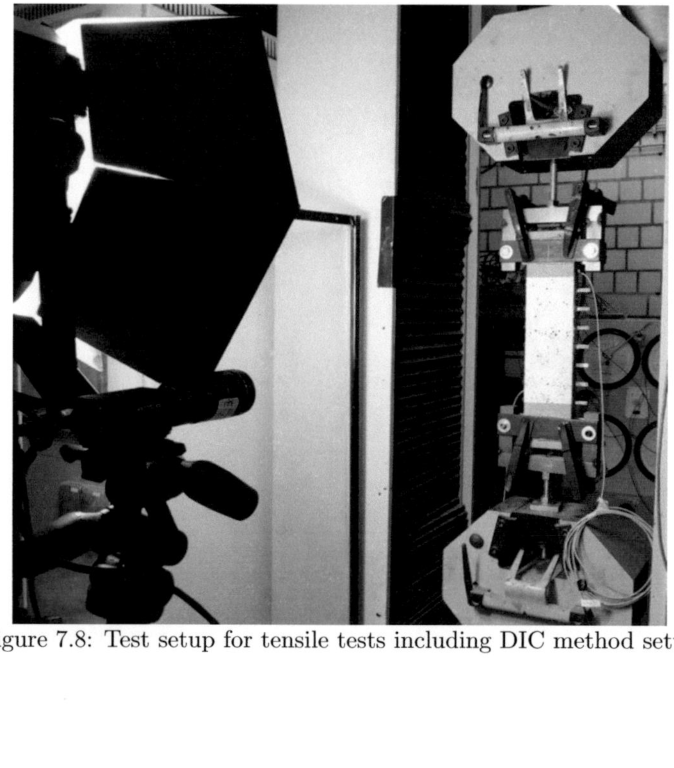

Figure 7.8: Test setup for tensile tests including DIC method setup

7.2.2 Tensile strength results

The typical pattern of failure for all tensile specimens was similar. Single cracks occurred in the proximity of the load introduction either on the upper or the lower side and in the middle area of the specimens. Usually, after three or four cracks appeared (Fig. 7.9), failure was observed as the existing cracks became wider and finally a pull-out of the reinforcement could be observed. Figures 7.10 - 7.12 present three example load-displacement and load-textile strains diagrams for specimens from each series, while diagrams for all tested specimens are included in Appendix C. Displacement at the diagrams is meant here as the machine path, load as measured with the load cell. Additionally, strains of the inner textile reinforcement in its middle point are read from the DFOS measurements. The strain readings from DFOS sensors were performed in intervals and therefore the measurements are not continuous. It must be though considered that square points on the diagrams mean the measured strains and the lines between measurements are only linear approximations of the load-strain relationship. The values of load, corresponding displacement and textile strains in the middle of the specimen achieved at the first cracking are summarized in Table 7.4.

Figure 7.9: Specimen during the test with two cracks visible

7.2.3 Coefficient of efficiency

A factor describing the efficiency of the reinforcement in concrete composite, and assessing the bond strength between textile reinforcement and concrete, is known as the coefficient of efficiency (COE) (Williams Portal, 2015; Voss, 2008). It is defined as the ratio between the maximum tensile strength of the

Table 7.4: Tensile tests results - load at first crack and corresponding displacement and textile strain

Specimen	Load at first crack [kN]	Displacement [mm]	Textile strain [με]
C00_T_a	12.97	1.76	78.8
C00_T_b	13.66	1.49	110.1
C00_T_c	13.32	1.33	99.7
C00_T_mean	**13.32** ± 0.34 (100%)	**1.53** ± 0.22 (100%)	**96.2** ± 15.9 (100%)
C15_T_a	15.63	1.82	138.6
C15_T_b	13.50	1.70	87.1
C15_T_c	17.86	2.22	125.0
C15_T_mean	**15.76** ± 2.18 (118%)	**1.91** ± 0.27 (125%)	**116.9** ± 26.7 (122%)
C20_T_a	20.58	2.66	202.8
C20_T_b	17.39	0.56	145.8
C20_T_c	16.15	2.67	149.9
C20_T_mean	**18.04** ± 2.29 (135%)	**2.57** ± 0.16 (168%)	**166.2** ± 31.8 (173%)

textile in the uniaxial tensile test to the tensile strength of the yarn in textile reinforcement itself.

$$COE = \frac{\sigma_{tu,max}}{\sigma_{y,u}} \tag{7.2}$$

where:
$\sigma_{y,u}$ is the tensile strength of one yarn, $\sigma_{\mathrm{y,u}} = 2500\ \mathrm{N/mm^2}$
$\sigma_{tu,max}$ is the maximum tensile strength of the textile in uniaxial tensile test defined as:

$$\sigma_{tu,max} = \frac{F_u}{A_t} \tag{7.3}$$

where:
F_u is the maximum load in uniaxial tensile test
A_t is the textile reinforcement cross-section, $\mathrm{A_t} = 16.26\ \mathrm{mm^2}$

If the COE value is higher than one, it implies good interaction between concrete and textile reinforcement and the tension stiffening effect, if it is

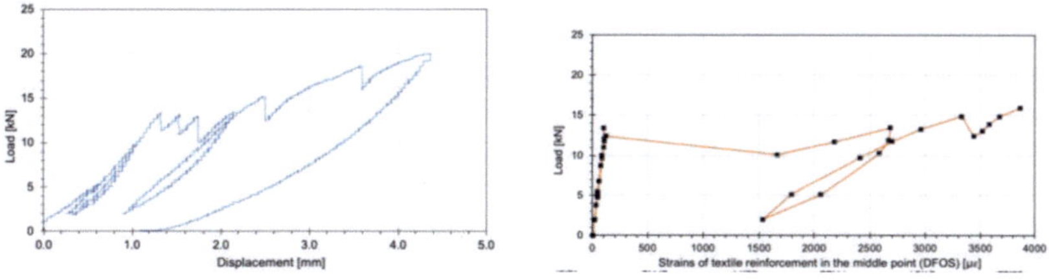

Figure 7.10: Load-displacement relationship for specimen C00_T_c and corresponding load-textile strain diagramm

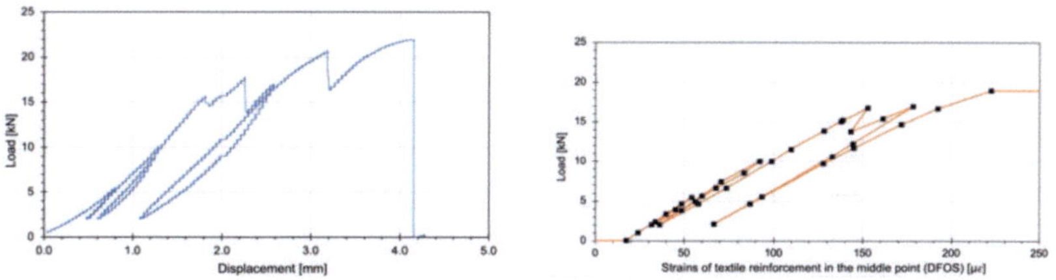

Figure 7.11: Load-displacement relationship for specimen C15_T_a and corresponding load-textile strain diagramm

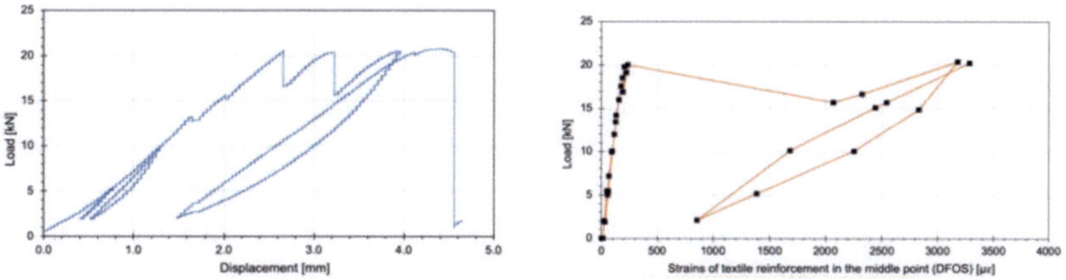

Figure 7.12: Load-displacement relationship for specimen C20_T_a and corresponding load-textile strain diagramm

lower than one, it means that the bond strength is low. For concrete mixtures C00, C15 and C20 these values are presented in Table 7.5.

Table 7.5: Load and stress at initial cracking and coefficient of efficiency for tensile specimens

	$F_{cr.T}$ [kN]	$\sigma_{cr.T}$ [N/mm^2]	COE [-]
C00_T_mean	13.32	4.44	0.33
C15_T_mean	15.76	5.23	0.39
C20_T_mean	18.04	6.01	0.44

7.2.4 Chemical prestress of tensile specimens

To estimate the prestress effect in the tensile specimens from the C20 series, the following calculations have been made. Firstly, tensile stress at cracking was calculated based on the results from Table 7.4. With concrete cross-section area assumed as $A_c = 3000$ mm^2, the stresses at initial cracking were calculated and are shown in Table 7.5.

According to Eq. 5.4, the ratio between direct tensile strength of non-reinforced C00 and C20 concretes is equal to $\alpha_{fctm} = 1.18$. Therefore it is here assumed, that without internal restraint, the tensile stress of C20 concrete specimens will be 1.18 times higher than of C00:

$$\sigma_{cr.T.C20.theor} = \sigma_{cr.T.C00} \cdot \alpha_{fctm} = 4.44\,N/mm^2 \cdot 1.18 = 5.24\,N/mm^2 \quad (7.4)$$

However, stress value obtained from experiments ($\sigma_{cr.T.C20}$) is higher than the calculated value ($\sigma_{cr.T.C20.theor}$). As the reinforcement ratio is equal in both specimens, this difference can be attributed to the influence of chemical prestress, which can be assessed directly as the difference between theoretical and experimental values:

$$\sigma_{pre.T.C20} = \sigma_{cr.T.C00} - \sigma_{cr.T.C20.theor} \quad (7.5)$$

$$\sigma_{pre.T.C20} = 6.01\,N/mm^2 - 5.24\,N/mm^2 = 0.77\,N/mm^2 \quad (7.6)$$

Prestress of these specimens could be calculated also basing on strain measurements. After 21 days, C00 specimens reached shrinkage with a magnitude of -0.367 mm/m (mean value), while C20 specimens achieved expansion of 0.410 mm/m (mean value). It corresponds with prestress in concrete equal to 0.75 N/mm^2 as described in Chapter 6, Eq. 6.2 - 6.5. This value calculated based on strains is nearly the same as the value calculated from the tensile test results and increase in cracking load.

7.2.5 Strains residuum

For all tensile specimens, the first and second loading and unloading cycles took place in an uncracked state. Therefore, linearly elastic behaviour could have been expected, and the loading and unloading curves should overlay with a very good agreement. It is indeed true for C00 specimens, however, already in the second loading cycle for C15 and C20 specimens we can observe that after second unloading, strain values do not fit exactly the value before the loading cycle. It implies that a strain residuum might be assumed. Therefore, the following analysis has been made: for all specimens, strains in the middle points of the DFOS sensors were read at the time points when a load of 2 kN was achieved (at first loading cycle and after the first two unloadings, see Fig. 7.13). It can be observed that the difference in strains of the textile reinforcement is the lowest for the reference C00 concrete, while for C15 and C20 specimens it increases. While for C15 the values have still a very high coefficient of variation, in series C20 a stable tendency is visible. Figure 7.14 present these differences. These observations match well with the phenomenon described in Chapter 3.4, which says that expansive mortars have greater deformability before cracking and the increase in load bearable before cracking can be contributed to both chemical prestress and partially to non-linear behaviour of expansive concrete described by Elasto-Plastic Fracture (EPF) model, described in Chapter 3.4.

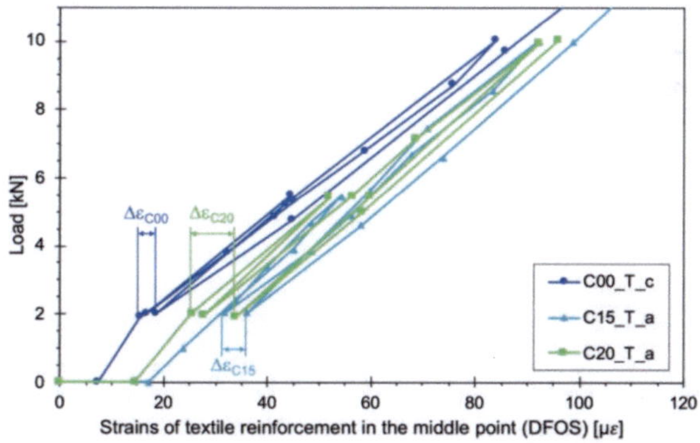

Figure 7.13: Load - strains diagrams of chosen tensile specimens, 1st and 2nd loading cycles and the strain residuum $\Delta\varepsilon_c$

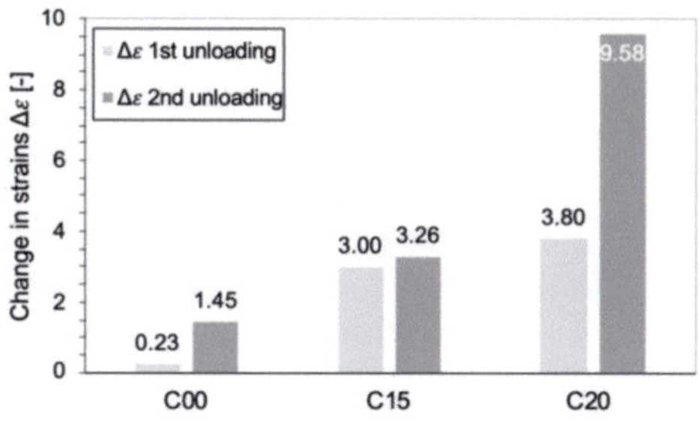

Figure 7.14: Strains residuum values in textile reinforcement after 1st and 2nd loading cycles

7.3 Summary

7.3.1 Summary of pull-out tests

Pull-out tests conducted on specimens made of C00 and C20 concrete shown, that specimens made of expansive concrete have similar behaviour in the pull-out tests, but achieve higher bond strength than reference specimens (5.99 N/mm^2 vs 7.46 N/mm^2). However, in the experimental programme presented here only three or four specimens were tested in each series. This number of specimens allows only a preliminary analysis, but there is a lack of more results that can be better statistically validated. Additional specimens are therefore required for a more reliable statement on the possible increase in the bond strength and its correlation with the expansion of concrete. Nevertheless, it is clear that expansive concrete will achieve at least the same or presumably even greater bond strength than the reference concrete without the addition of an expansive admixture.

7.3.2 Summary of tensile tests

Tensile tests were conducted on specimens made of reference C00 and expansive C15 and C20 concretes, all specimens were reinforced with carbon textile reinforcement. In tensile tests, similar cracking pattern was observed for each element, consisting of three or four cracks over the length of 300 mm between clamping plates. Load at first crack increased for specimens made of expansive concrete in comparison with reference specimens by 18% for C15 and 35% for C20 concrete. Furthermore, optical sensors fixed on the internal textile reinforcement allowed to measure textile strains during tensile tests, these strains at the moment of the first crack also increased (by 22% for C15 and 73% for C20 specimens).

Besides the visible improvement in cracking load, a so-called strain residuum phenomenon was observed for tensile specimens. It means that after several loading and unloading cycles even in the theoretically elastic state of a linear stress-strain relationship, some deformation remain in the expansive concrete specimens. This strain residuum becomes higher with increasing expansive admixture amount.

Because of the uniaxial stress state, the prestress introduced to the specimens could have been assessed analytically. In C20 elements the concrete prestress achieved 0.77 N/mm^2. This value of prestress calculated based on the tensile tests correlates very well with value based on the strain measurements (0.77 N/mm^2 vs 0.75 N/mm^2).

Chapter 8

Flexural tests

8.1 Beam specimens

8.1.1 Test setup

Beam specimens were cast in addition to the non-load-bearing surface elements (slabs) which primarily bear only their own weight or resist horizontal wind loads as vertical panels. Beams were necessary because, due to the concrete cover required for corrosion protection, thin-walled slabs with steel mesh reinforcement which would be comparable to thin-walled slabs with textile reinforcement cannot be produced. Also, reference specimens with uniaxial reinforcement are advantageous to address the research objectives. Therefore, beam elements were cast and an axial reinforcement was inserted in the middle of the cross-section, so that the influence of the chemical prestressing became clearer since influences such as the involvement of the transversal reinforcement were excluded. However, the reinforcement ratio (as described in Chapter 4.2) was held constant with the reinforcement ratio in slabs with textile reinforcement.

The static four-point bending tests were carried out in the universal testing machine. The beams with a span of 1800 mm were placed under the frame with an attached servo-hydraulic jack. The distances between the load introduction points and the supports were 600 mm each. The test setup is shown schematically in Fig. 8.1 and in Fig. 8.2 in the real execution. A load cell with a measuring range of 50 kN was installed on the test cylinder. For the deflection measurements, two laser distance sensors with a measuring range of 40 mm were arranged in the middle of the beam on its bottom side, see Figures 8.2 and 8.3. In addition, the strain measurements on the reinforcement and in the concrete cross-section were carried out with fibre optic sensors and strain gauges described before. Also, Digital Image Correlation (DIC) method was used to capture the crack pattern in the central beam area (over a length of 860 mm) using the setup shown in Figure 8.4.

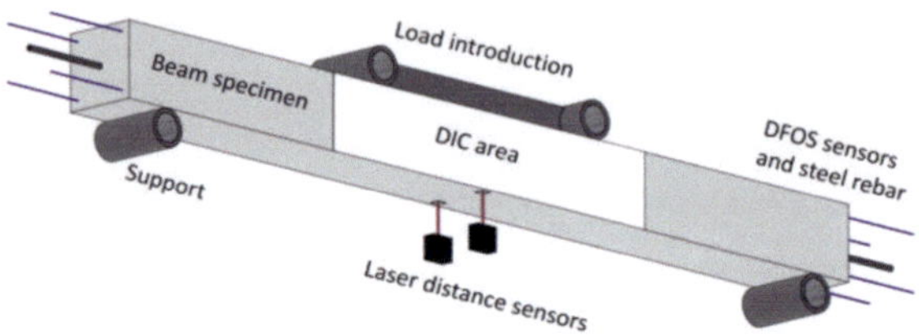

Figure 8.1: Presentation of the test setup

Figure 8.2: Test setup for four-point bending of a concrete beam with steel reinforcement

The load was applied in three loading and unloading cycles: first in a linear-elastic area, second after initial crack formation and third before maximum deflection was reached. For the first range, the load was increased at a constant speed $v_1 = 0.5$ mm/min up to a force of approx. 6.0 kN and then reduced again to approx. 1.0 kN. In the subsequent second loading cycle, the same loading speed was applied until crack formation was initiated and the load was then removed at a deflection of approx. 5.5 mm. The third and final loading phase with a loading speed of $v_2 = 2.5$ mm/min was completed at a deflection of between 14 mm and 16 mm and then final unloading took place.

Figure 8.3: Laser distance sensors for measuring the deflection of the beam in the middle

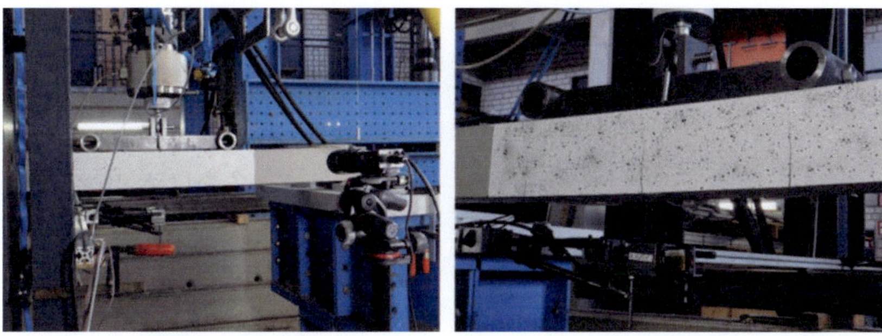

Figure 8.4: DIC method for capturing the crack pattern: camera with the measuring field (left) and typical crack pattern (right)

8.1.2 Load-displacement relationship

Figure 8.5 show the load-deflection diagrams of the four-point bending tests for selected test specimens, with the corresponding values and percentage differences summarized in Table 8.1. Single results for all beams are presented in Appendix B. Generally, the load responses are comparable for all specimens within one series. Apparently, the load that can be borne up to the initial crack formation increases with the increase of the expansive admixture amount. For the test specimens from the C20 series, the load at initial crack formation is on average 66% higher than the value for the reference C00 elements, the corresponding deflections differ by 76%. For the beams from the

C15 series, the load at initial crack formation is 14% higher and the deflection 7% higher compared to the reference elements. Even taking into account an 18% higher bending tensile strength of the test specimens from the C20 series compared to the reference C00 series, it can be concluded that prestress was introduced by expansive concrete.

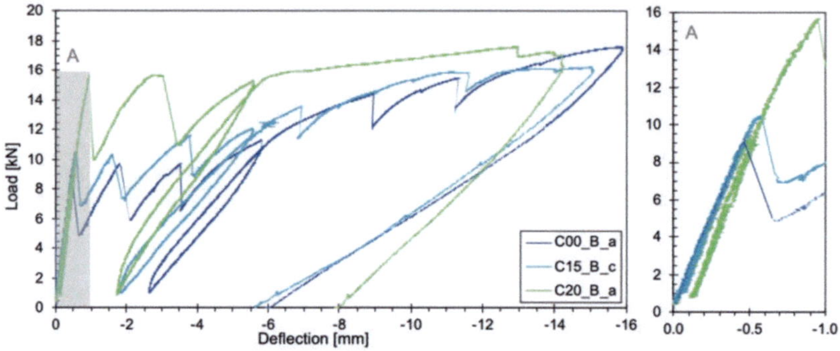

Figure 8.5: Load-deflection diagram for specimens from all series - comparison of chosen specimens: B_C00_a, B_C15_c, B_C20_a (left) and enlarged view until the first crack (right)

Table 8.1: Load and deflection at initial crack formation in beams subjected to flexure

Specimen name	Load at initial cracking [kN]	Corresponding deflection [mm]
C00_B_a	9.31	0.47
C00_B_b	10.12	0.55
C00_B_c	8.49	0.57
C00_B_mean	**9.30** ± 0.81 (100 %)	**0.53** ± 0.06 (100 %)
C15_B_a	11.85	0.67
C15_B_b	9.50	0.45
C15_B_c	10.50	0.58
C15_B_mean	**10.61** ± 1.18 (114 %)	**0.57** ± 0.11 (107 %)
C20_B_a	15.65	0.95
C20_B_b	14.25	0.84
C20_B_c	16.43	1.01
C20_B_mean	**15.45** ± 1.11 (166 %)	**0.93** ± 0.09 (176 %)

8.1.3 Cracking pattern

During the experiments, strains in concrete and on the rebar were recorded with DFOS, and on the rebar additional measurements with strain gauges were taken. The results of the strain measurements on the steel reinforcement bars are shown in the following diagrams (Figures 8.6 to 8.8 for three chosen beams). The diagrams on the left side show the strain curves over the entire length of the rebar. A new line is shown for each load level. Thanks to the sensor technology, it was possible to predict the cracks visible from the outside on the beam surface in advance based on the strain patterns. Diagrams on the right side show the strain values measured with strain gauges in time together with the load regime in the time of the experiment. Here single cracks are visible as sudden changes of strains, however their magnitude depends on how close to a strain gauge a crack appeared. No quantitative analysis is possible here. Also, contrary to the DFOS method, the cracks located further away from the strain gauges could not be directly detected.

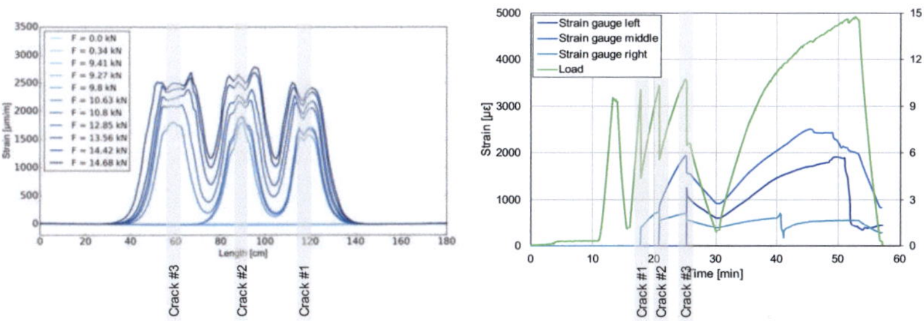

Figure 8.6: Strains of steel reinforcing bar along the beam B_C00_b for subsequent load steps, measured with DFOS

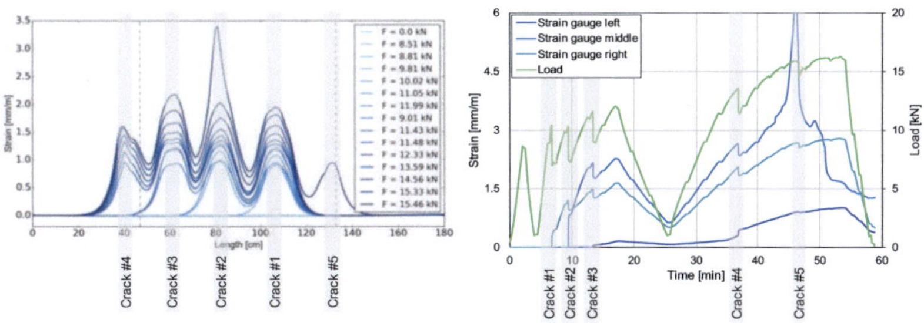

Figure 8.7: Strains of steel reinforcing bar along the beam B_C15_c for subsequent load steps, measured with DFOS

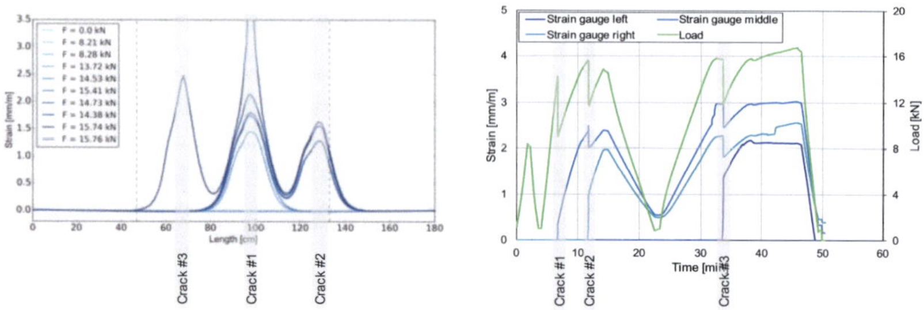

Figure 8.8: Strains of steel reinforcing bar along the beam B_C20_a for subsequent load steps, measured with DFOS

Finally, these strain measurement results could be well and precisely confirmed for higher load steps with the observation of cracks and digital image correlation (DIC) method (Fig. 8.9). Here however some of the cracks occurred outside of the DIC area (that is outside of the middle 86 cm of the 200-cm long beam).

Figure 8.9: Photo and results of DIC of beams B_C15_c and B_C20_a

For one chosen load step, strains within the element can be analysed. The strain values in the tensile area in the reinforcing steel layer (L0) and in the concrete (E5) are shown with the strains in the compression area (E1 and E2) on Figure 8.10. Strains in sensors on the bottom side (E3 and E4) are neglected as for the chosen load level they already reached their maximum range. The prominent peaks in the strain curves (especially E5) result from crack formation in the concrete and designate the crack locations. Due to the bond between concrete and steel and the development length, the strain curve for the reinforcing steel in the area of these peaks is typically smoothen. The location of the cracks can also be identified on the upper side of the element in the measured values of the sensors E1 and E2. The concrete compressive strains reach a magnitude of 0.5 mm/m in the areas of cracks.

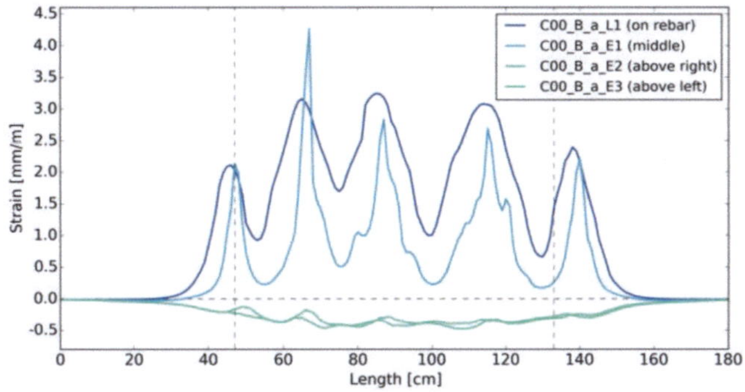

Figure 8.10: Strains measured in sensors E1, E2, E5 and L0 along the beam B_C00_a, peaks in strains designate positions of cracks

8.1.4 Chemical prestress of beam specimens

In order to estimate the prestress effect in the beam specimens from the C20 series, the cracking moment was calculated based on the results from Table 8.1. In four point bending test the distance between support and the constant moment area is equal to 0.6 m and the average cracking moments are shown in Table 8.2.

Table 8.2: Mean cracking loads and moments for beam specimens

	Cracking load [kN]	Cracking moment [kNm]
C00_B_mean	9.30	2.79
C15_B_mean	10.61	3.18
C20_B_mean	15.45	4.64

According to equation 5.4, the ratio between direct tensile strength of non-reinforced C00 and C20 concretes is equal to $\alpha_{\mathrm{fctm}} = 1.18$. Therefore it is here assumed, that without restrain in form of the internal reinforcement, the cracking moment of the C20 concrete specimens will be 1.18 times higher than of C00:

$$M_{cr.B.C20.theor} = M_{cr.B.C00} \cdot \alpha_f ctm = 2.79\, kNm \cdot 1.18 = 3.29\, kNm \quad (8.1)$$

However, cracking moment value obtained from experiments ($\mathrm{M_{cr.B.C20}} =$ 4.64 kNm) is higher than the calculated value ($\mathrm{M_{cr.B.C20.theor}} =$ 3.29 kNm). As the reinforcement ratio is equal in both specimens, this difference can be attributed to the influence of chemical prestress, which can be assessed in the following way:

$$M_{cr.B.C20} = M_{cr.B.C20.theor} + \frac{N_{C20}}{A_c} \cdot W_y \tag{8.2}$$

$$N_{C20} = \frac{1350000\,Nmm \cdot 25600\,mm^2}{682666.7\,mm^3} = 50.63\,kN \tag{8.3}$$

$$\sigma_{pre.B.C20} = \frac{N_{C20}}{A_c} = \frac{50630\,N}{25600\,mm^2} = 1.98\,N/mm^2 \tag{8.4}$$

With these assumptions beams made of C20 concrete are effectively pre-stressed with stress equal to 1.98 N/mm^2.

8.2 Mid-scale slabs with textile reinforcement

8.2.1 Test setup

In mid-scale flexural test specimens described in 4.3.2 were subjected to four-point bending (Fig. 8.11). The distance between supports equalled 1000 cm, and the distance between support and load 400 mm (Fig. 8.12). Tests were conducted in an electromechanical testing machine equipped with a load cell with a 25 kN measuring range. The tests were carried out in a displacement-controlled manner, whereby the loading speed was 1 mm/min until the first crack and 5 mm/min until failure. During bending, the force and the deflection in the middle of the plate on both sides were measured with a frequency of 10 Hz. The maximum measuring range of the inductive displacement transducers (LVDT) used was 40 mm, after this value was reached further displacement was recorded by the machine path (Zdanowicz and Marx, 2018*b*).

Figure 8.11: Four-point bending test of a mid-scale slab with carbon textile reinforcement

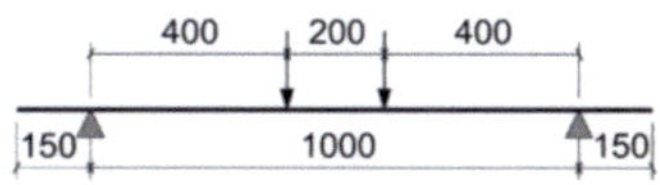

Figure 8.12: Test setup for the mid-scale flexural tests

8.2.2 Load-displacement relationship

A typical load-displacement diagram for a C00 series of specimens is shown in Figure 8.13. The crack formation is visible in the diagram as sudden drops in the load. The load at initial crack formation increases with an increasing amount of the expansive admixture. The same applies to the corresponding deflection during initial crack formation. The results of the load and deflection measurements at initial cracking are summarized for all samples in Table 8.3. Table 8.4 contains the corresponding test data for the state of failure. Noteworthy, load at initial cracking for specimens from the C15_F and C20_F series is somehow similar, however the corresponding deflection is higher for C20_F series. The ultimate state of bending moment should theoretically not be influenced by chemical prestressing(Japan Society of Civil Engineering, 1994), however it was observed that maximum load also increased with an increase in expansive admixture amount. The differences were however not high and the maximum deformation was similar for all specimens. Figure 8.14 shows the load-deflection diagram for selected slabs. These curves are shown in the enlarged form up to the first crack for the area up to a deflection of approx. 4 mm.

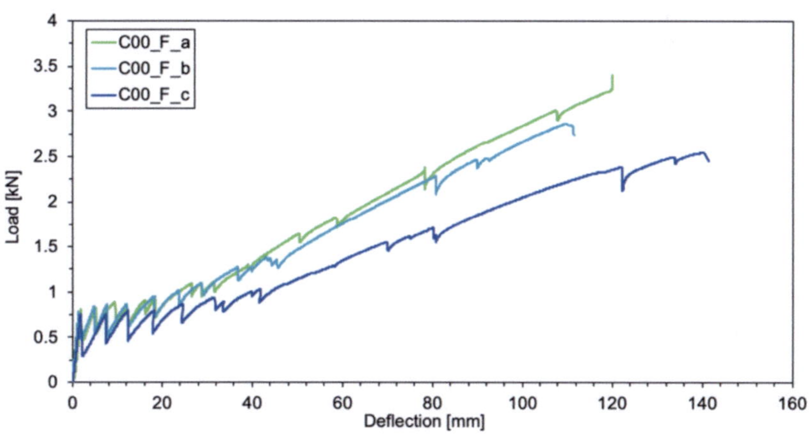

Figure 8.13: Load-displacement path for three specimens of C00 series

Table 8.3: Load and deflection at initial crack formation

Specimen name	Load at initial crack formation [kN]	Deflection at initial crack formation [mm]
C00_F_a	0.80	0.67
C00_F_b	0.76	1.05
C00_F_c	0.78	1.37
C00_F_mean	**0.78** ± 0.02 (100 %)	**1.03** ± 0.35 (100 %)
C15_F_a	1.01	1.38
C15_F_b	0.99	1.66
C15_F_c	0.91	1.52
C15_F_mean	**0.97** ± 0.05 (124 %)	**1.52** ± 0.14 (148 %)
C20_F_a	1.01	1.78
C20_F_b	0.94	1.81
C20_F_c	0.99	2.04
C20_F_mean	**0.98** ± 0.04 (126 %)	**1.88** ± 0.14 (182 %)
C25_F_a	1.26	1.90
C25_F_b	1.25	2.50
~~C25_F_c~~*	~~0.95~~*	~~2.19~~*
C25_F_mean	**1.25** ± 0.01 (161 %)	**2.20** ± 0.30 (213 %)

* - Test specimen *C25_F_c* was excluded from the evaluation due to premature failure.

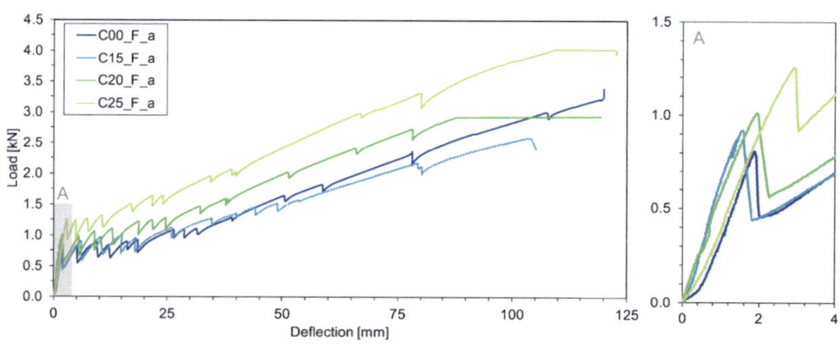

Figure 8.14: Load-deflection diagram for all mid-scale slabs and an enlarged view of initial 4 mm deflection

8.2.3 Cracking pattern and crack spacing

Figure 8.15 shows a typical closed crack pattern for the slabs. The number of cracks was comparable for all test specimens in a closed crack pattern (between 11 and 13) and independent of the concrete mixture. However, the crack spacing for the individual series varied: C00 - 67.0 mm, C15 - 68.0 mm, C20 - 74.1 mm and C25 - 72.6 mm. This means that for the C20 and C25 series, the crack spacing is increased by 10% and 8% respectively compared to the reference series C00.

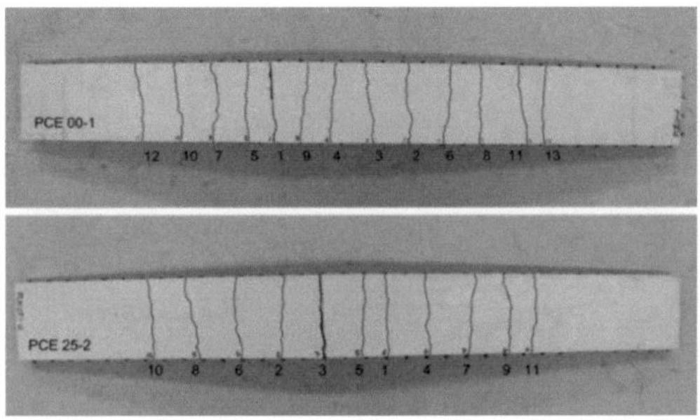

Figure 8.15: Typical crack pattern for the mid-scale slabs from series C00 and C25

Table 8.4: Load and deflection at failure

Specimen name	Load at failure [kN]	Deflection at failure [mm]
C00_F_a	3.41	129
C00_F_b	2.87	113
C00_F_c	2.55	137
C00_F_mean	**2.94** ± 0.44 (100 %)	**126** ± 12 (100 %)
C15_F_a	2.59	101
C15_F_b	2.72	107
C15_F_c	3.01	126
C15_F_mean	**2.78** ± 0.22 (94 %)	**112** ± 13 (88 %)
C20_F_a	3.91	133
C20_F_b	3.06	134
C20_F_c	3.16	127
C20_F_mean	**3.38** ± 0.47 (115 %)	**131** ± 4 (104 %)
C25_F_a	4.03	122
C25_F_b	4.25	141
~~C25_F_c~~*	~~2.89~~*	~~107~~*
C25_F_mean	**4.14** ± 0.16 (141 %)	**131** ± 14 (104 %)

* - Test specimen *C25_F_c* was excluded from the evaluation due to premature failure.

Higher crack spacing of chemically prestressed specimens was confirmed also in the literature. It can be attributed to a higher tension stiffening effect of chemically prestressed members. "Larger distance between two adjacent cracks reduces number of cracks and thus the reduction of average stress caused by stress redistribution at crackings" (Sahamitmongkol and Kishi, 2011). However, this effect is strongly dependent on the provided concrete cover, because the thickness of concrete might affect the stress transfer from concrete to the internal reinforcement.

8.3 Large-scale slab specimens

8.3.1 Test setup

Large-scale slab specimens with dimensions of 2000 x 1000 x 30 mm^3 (described in chapter 4.3.2) were casted as main test specimens in the thesis work. They were reinforced with carbon textile reinforcement located in the middle of their depth. Slabs were subjected to static four-point bending tests in the universal testing machine. The slabs were placed with a span of 1800 mm under the frame with an attached servo-hydraulic jack. The distances between the load introduction and the supports were 600 mm each. The test setup is shown schematically in Fig. 8.16 and in Fig. 8.17 in the real execution.

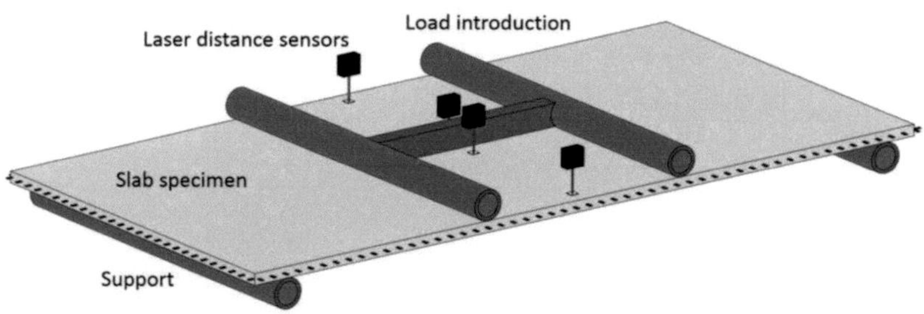

Figure 8.16: Presentation of the test setup

A load cell with a measuring range of up to 50 kN was installed on the test cylinder. For the deflection measurements, four laser distance sensors with a measuring range of 40 mm were arranged in the middle line of the slab's span, see Figures 8.17 and 8.18. In addition, strain measurements on the reinforcement and in concrete were carried out with distributed fibre optic sensors described in chapter 6.1.4.

Figure 8.17: Test setup for four-point bending of a concrete slab with carbon textile reinforcement

Also, the Digital Image Correlation (DIC) method was used to capture the crack pattern in the central slab area (on a field with dimensions of 620 x 860 mm^2) using the setup shown in Fig. 8.18. Here a machine vision camera (type Manta G-235 from Allied Vision) was used and photos were made every two seconds. All measuring devices were synchronised together so that the timing of the photos shot can be directly related to the moment of testing machine readings or DFOS measurements.

Figure 8.18: Laser distance sensors for measuring the deflection of the slab in the middle (left top), preparation of the DIC area (left bottom) and DIC setup for capturing the crack pattern (right)

A loading regime with five loading and unloading cycles was applied. For the first two load cycles a maximum force of approx. 1.4 kN was applied at a constant speed $v_1 = 0.5$ mm/min. These two loading cycles should provide a record of the load-deflection curves in the still substantially linear strain range. Each unloading was realized to approx. 0.2 kN. Further, a third cycle was performed up to a load level when deflection of about 5 mm occurred and released again to 0.2 kN. Then, the load speed was increased to $v_2 =$ 2.5 mm/min and the fourth load cycle was completed at a load for which corresponding deflection achieved approx. 25 mm. The fifth load cycle was run at a load speed of $v_3 = 10$ mm/min up to a deflection of 36 mm, which corresponded to 1/50 of the span length. A load of approximately 1.5 kN - 3.5 kN was applied at this deflection level. Finally, the slabs were unloaded.

8.3.2 Load-displacement relationship

In each of the three diagrams in Figure 8.19, the formation of single cracks is visible as a drop in the load. The initially linear increase in load is followed by a phase of crack formation, which is accompanied by an increase in deflection. The slope of the load-deflection curves in the linear non-cracked area is approximately the same for all examined test specimens since it reflects the plate stiffness of the non-cracked cross-section. However, the curves differ for both the uncracked and the cracked cross-section in the magnitude of the load that can be carried. Diagrams for all slabs are included in Appendix A.

The force-deflection curves show that all test specimens with the addition of expansive admixture can withstand higher loads during crack formation than the test specimens from the reference concrete. The results of the flexural tests for all test specimens are summarized in Table 8.5. The first crack of the test specimen C20_S_a was difficult to detect since no corresponding force drop was observed in the force-deflection curve. The crack formation and its development could be thus detected by means of the Digital Image Correlation method as well as with the measurements acquired from the fibre optic sensors. This will be discussed in the following section.

Furthermore, strains of concrete in the longitudinal direction were measured on the top and bottom surfaces of the slabs with two 100 mm long electrical strain gauges. Strain values in reference to the load applied are presented in the Figure 8.20. The strains measured with the strain gauges located on the top surface were comparable for all specimens and reached between 0.5 - 0.75 mm/m. On the bottom surface, strains were measured until a crack occurred which went through the strain gauge, from this moment a linear increase of measured strains without changes in load is observed and measurement results are no longer relevant. It is visible that in the second load cycle, when deflection reached about 5 - 6 mm, strains were much higher in C00 and

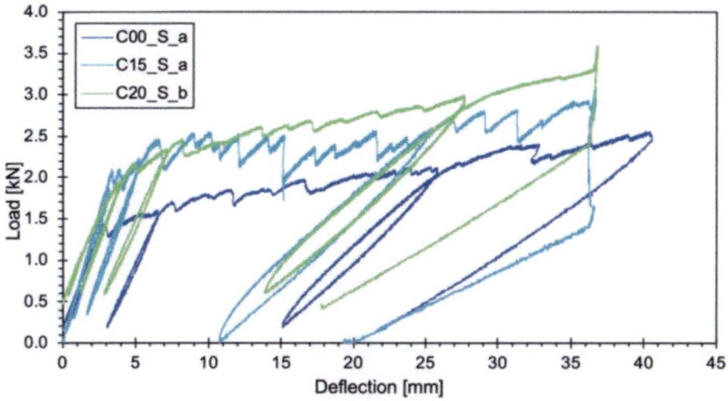

Figure 8.19: Comparison of selected slabs from series C00, C15 and C20 in terms of their load-deflection relationship

Table 8.5: Load and deflection at initial crack formation in slabs

Specimen name	Load at initial cracking [kN]	Corresponding deflection [mm]
C00_S_a	1.46	2.97
~~C00_S_b~~*	~~0.33~~*	~~2.66~~*
C00_S_c	1.74	2.69
C00_S_mean	**1.60** ± 0.20 (100 %)	**2.83** ± 0.20 (100 %)
C15_S_a	2.05	3.41
C15_S_b	2.89	3.17
C15_S_c	2.39	3.81
C15_S_mean	**2.44** ± 0.42 (153 %)	**3.46** ± 0.32 (122 %)
~~C20_S_a~~*	~~1.62~~*	~~6.95~~*
C20_S_b	2.16	5.02
C20_S_mean	**2.16** ± 0.00 (135 %)	**5.02** ± 0.00 (177 %)

* - In the case of slab C20_S_a, no cracks that could be identified by a drop in force were visible in the force-deflection diagram. The values in the table therefore refer to the cracks visible in the DIC results.

C15 specimens than in C20 slab (0.75 or 1 mm/m in comparison with 0.25 mm/m).

On the left side of the diagrams, in the compresive zone, there is a difference visible between reference C00 mixture and expansive concrete mixtures, C15 and C20. For C00, at a maximum load of 2.5 kN strains in the strain gauge at the top increase but no additional load can be absorbed. On the contrary, in C15 and C20 specimens an increase of strains is visible together with an increase of load until the end of measurements. Thus, a prestressing effect might be anticipated by observing the compressive zone, as in chemically prestressed specimens additional compression is introduced in the specimens which allow for higher load levels.

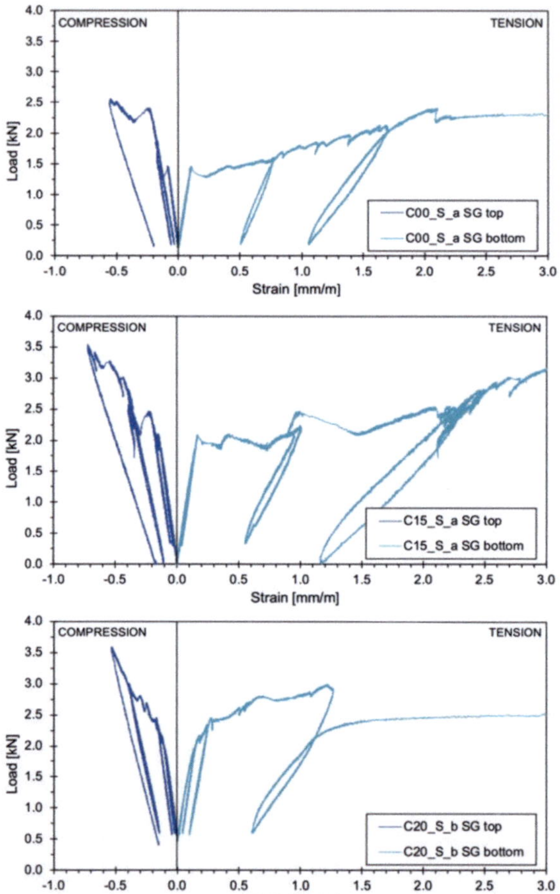

Figure 8.20: Strains in concrete on the top and bottom surface for the slabs C00_S_a (top), C15_S_a (middle) and C20_S_c (bottom)

8.3.3 Cracking pattern

During the four-point bending tests, the strains in the concrete and on the textile reinforcement were examined with distributed fibre optic sensors. Results of these measurements for one slab from each series are presented in Figures 8.21 to 8.23 - strain peaks for each load level correlate with crack locations. A fibre optic sensor that was embedded in concrete in the middle of the slab in the longitudinal direction is considered here. These diagrams illustrate both where cracks occur and how they develop with increasing load. In addition, the measured values of the fibre optic sensors can be used to detect cracks at an early stage even if they are not yet visible.

These findings correspond also very good with the crack development in the middle of the slabs recorded by DIC method (Fig. 8.24). DIC analysis was performed with GOM Correlate software using square facets with a facet size of 11 px (to aim for better acquisition of local effect within the facet) and a point distance of 9 px (to provide overlapping area and higher resolution). Deformations in the longitudinal direction are presented here.

Figures 8.25 to 8.27 show the correlated results of the fibre optic measurements and the DIC analysis for one chosen load level designated on the load-displacement diagram. Vertical lines designate the width of the DIC area. DFOS measurements allow thus a very good prediction of the crack locations, even before they can be observed with the DIC. Further results of the measurements for all slabs are shown in Appendix A.

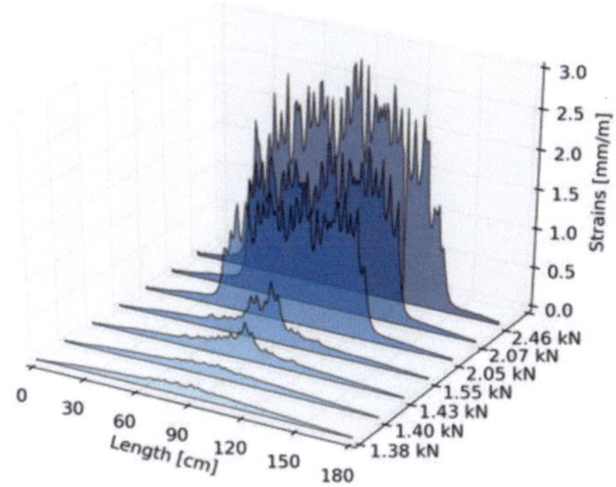

Figure 8.21: DFOS measurements - cracking pattern for the slab C00_S_a

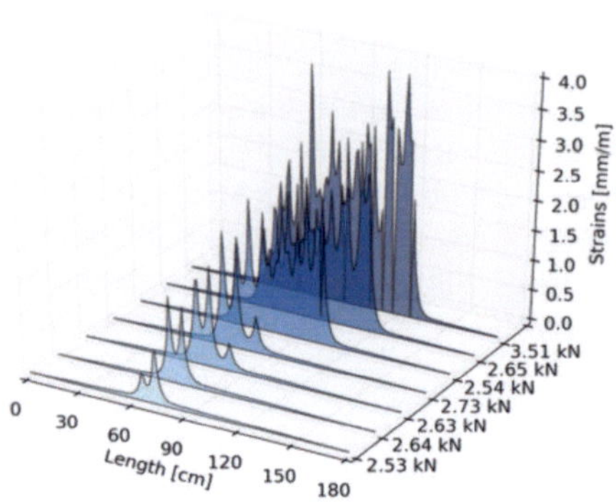

Figure 8.22: DFOS measurements - cracking pattern for the slab C15_S_b

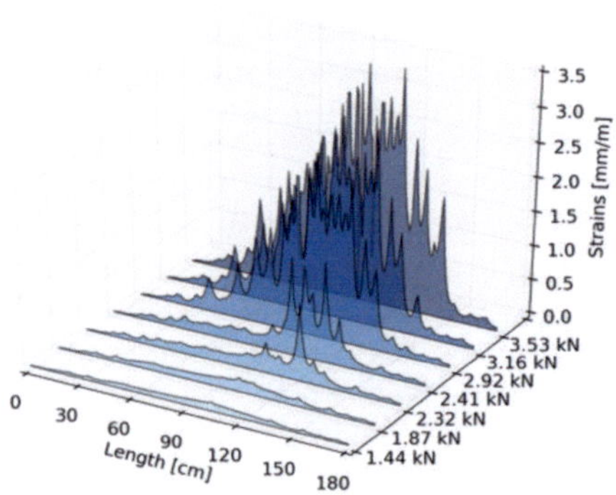

Figure 8.23: DFOS measurements - cracking pattern for the slab C20_S_b

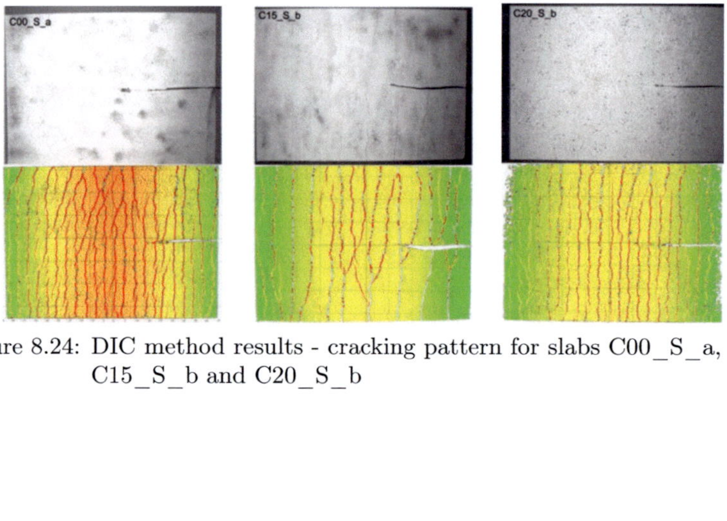

Figure 8.24: DIC method results - cracking pattern for slabs C00_S_a, C15_S_b and C20_S_b

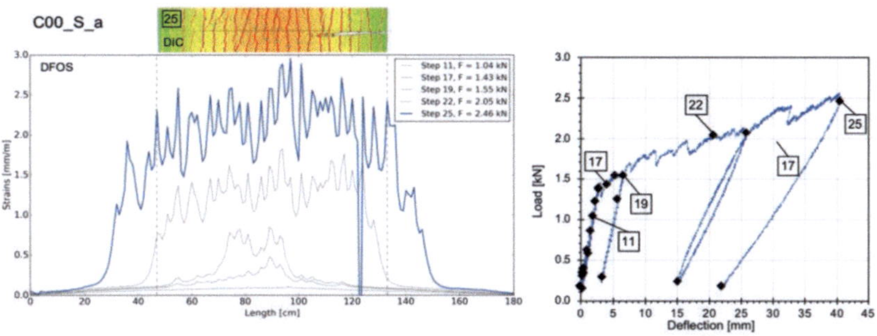

Figure 8.25: Comparison of measurements results from DFOS sensors with DIC method for the slab C00_S_a, load step 25

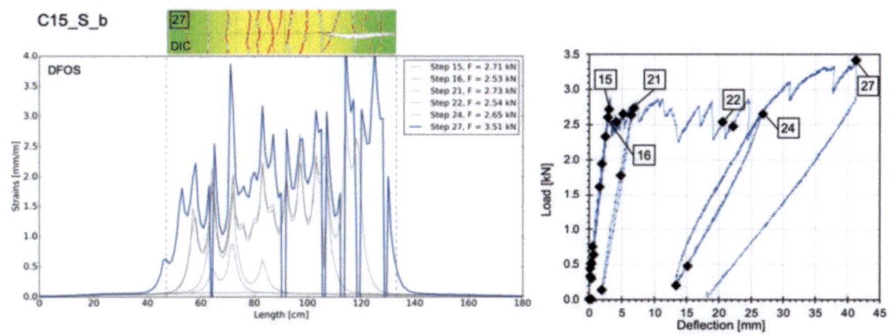

Figure 8.26: Comparison of measurements results from DFOS sensors with DIC method for the slab C15_S_b, load step 27

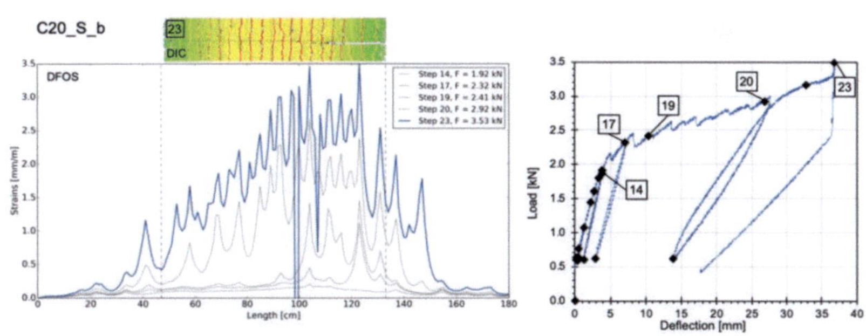

Figure 8.27: Comparison of measurements results from DFOS sensors with DIC method for the slab C20_S_b, load step 23

Further observation was made regarding the damages in the concrete structure even before the first crack occurred. Strain distribution along slabs is shown for specimens C00_S_a and C20_S_c in Figure 8.29 for chosen load steps. It can be observed, that even in the theoretically elastic state (before cracking), some local damages can be observed in concrete (load step 13 for C00_S_a and 14 for C20_S_c). After subsequent unloading, the deflection of slabs goes back to the state before loading, but these damages to some extent remain in concrete - the damaged area remains similar, although peak values decrease (load step 14 for C00_S_a and 15 for C20_S_c). There were no cracks visible at this time on DIC results neither a drop in force on the load-deflection diagram.

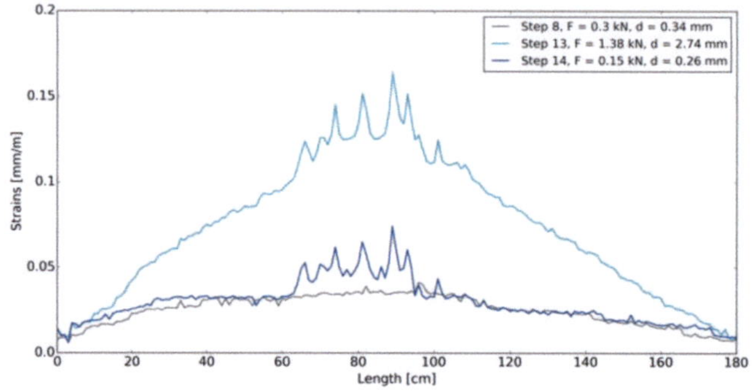

Figure 8.28: Results from DFOS sensors after unloading cycle for slab C00_S_a

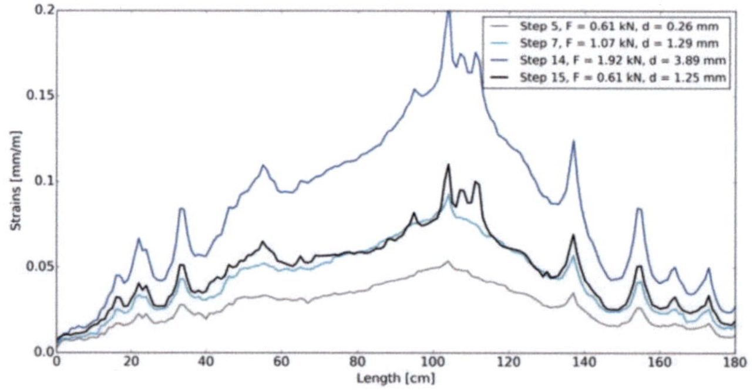

Figure 8.29: Results from DFOS sensors after unloading cycle for slab C20_S_b

Also when strains in concrete (free DFOS fibres) and on the textile reinforcement (glued DFOS fibres) were compared, a difference was visible on every loading stages (Figure 8.30). Although strain peaks occur in the same locations of cracks, strains of textile reinforcement (light blue lines) are lower than strains in concrete (dark blue lines).

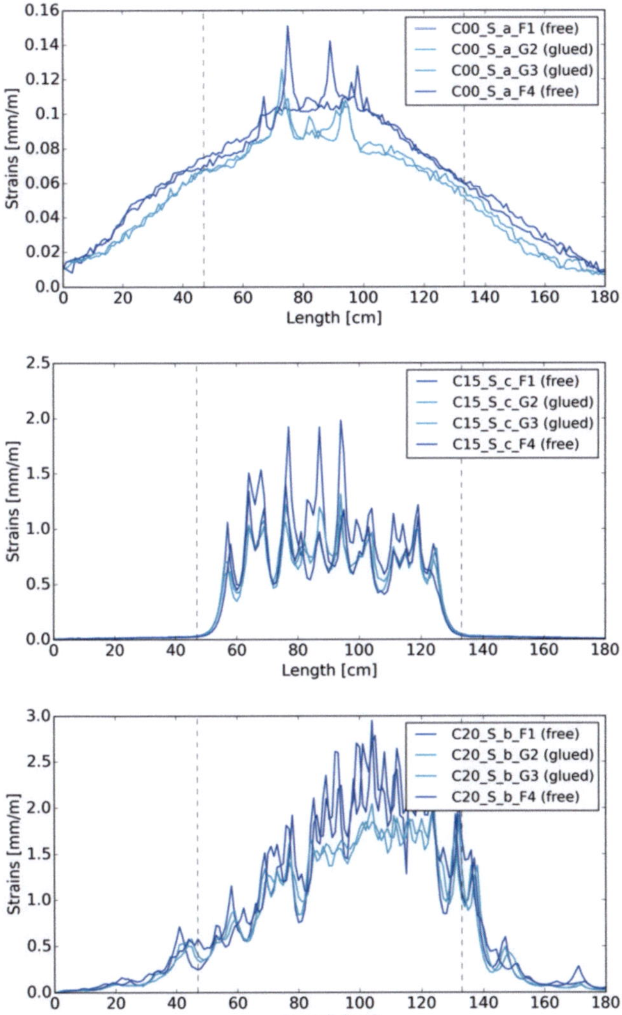

Figure 8.30: Comparison of strains measured with glued and free DFOS sensors for slabs C00_S_a (load step 13), C15_S_c (load step 23) and C20_S_b (load step 22)

8.3.4 Analysis of prestress

With an aim to analyse the prestress introduced with the expansion of concrete, cracking moments of C00, C15 and C20 slabs were calculated and compared. Firstly, the cracking moment from experimental results was calculated with $\mathrm{F_{cr}}$ values taken as mean values from Table 8.5 and the average cracking moments are shown in Table 8.6.

Table 8.6: Mean cracking load and moments for slab specimens

	Cracking load [kN]	Cracking moment [kNm]
C00_S_mean	1.60	0.48
C15_S_mean	2.44	0.73
C20_S_mean	2.16	0.65

Subsequently, a theoretical values of cracking moments were set for C00 and C20 slabs considering concrete tensile strength with the following formula:

$$M_{cr} = (f_{ctm} + \frac{N}{A_c}) \cdot W_y \tag{8.5}$$

For plain concrete without reinforcement the direct tensile strength value ($\mathrm{f_{ctm}}$) can be used as calulated before in Eq. 5.2. As the slabs are reinforced with textile reinforcement, the following assumptions are made after (Kulas, 2013):

$$F_{cr} = F_{cr.c} + F_{cr.t} \tag{8.6}$$

where $F_{cr.c}$ is the tensile force in concrete at cracking $F_{cr.t}$ is the tensile force in textile reinforcement at cracking. It is also assumed, that in phase I a linear stress-strain state is provided, the reinforcement area is neglected ($A = A_c$) and a full bond between concrete and reinforcement is provided so that $\epsilon_{cr.c} = \epsilon_{cr.t}$. Thus, the compound element tensile strength at first crack can be calculated with mechanical reinforcement ratio based on stiffness, ω_{l}.

$$\omega_l = \frac{E_t \cdot A_t}{E_c \cdot A_c} \tag{8.7}$$

For slab made of C00 concrete, the mechanical reinforcement ratio is:

$$\omega_{l.C00} = \frac{180000 \cdot 142}{33420 \cdot 29858} = 0.026 \tag{8.8}$$

and the $\mathrm{f_{ctm.s.C00}}$:

$$f_{ctm.s.C00} = f_{ctm.C00} \cdot (1 + \omega_{l.C00}) = \tag{8.9}$$

$$= 3.13\,N/mm^2 \cdot (1 + 0.026) = 3.21\,N/mm^2 \tag{8.10}$$

With the $f_{ctm.s}$ value calculated in Eq. 8.10 and the N value set to zero and substituted to Eq. 8.5, a following theoretical value of cracking moment is obtained for C00 specimens:

$$M_{cr.s.C00} = 3.21\,N/mm^2 \cdot \frac{1000\,mm \cdot (30\,mm)^2}{6} = 0.48\,kNm \tag{8.11}$$

For the C00 concrete, the theoretical value shows a very good agreement with the value from experimental tests (0.48 kNm in both cases).

The calculations for C20 concrete takes the following form:

$$\omega_{l.C20} = \frac{180000 \cdot 142}{32290 \cdot 29858} = 0.027 \tag{8.12}$$

$$f_{ctm.s.C20} = 3.70\,N/mm^2 \cdot (1 + 0.027) = 3.80\,N/mm^2 \tag{8.13}$$

$$M_{cr.s.C20} = (3.80\,N/mm^2 + \frac{N_{C20}}{30 \cdot 1000\,mm^2}) \cdot \frac{1000\,mm \cdot (30\,mm)^2}{6} \tag{8.14}$$

As in Eq. 8.14 the M_{cr} and N values are unknown and a good agreement between theoretical and experimental values was observed for C00 slabs, here the cracking moment M_{cr} will be assumed as equal to the value from experiments, i.e. 0.65 kNm:

$$N_{C20} = 26.7\,kN \tag{8.15}$$

In C20 concrete, a prestressing force of 26.7 kN can be therefore assumed, which corresponds with the prestress value of:

$$\sigma_{pre.slab.C20} = \frac{N_{C20}}{A_c} = \frac{26700\,N}{30000\,mm^2} = 0.89\,N/mm^2 \tag{8.16}$$

Finally, a theoretical ultimate load-bearing capacity is calculated below according to (Voss, 2008; Williams Portal, 2015). The aim is to assess whether the cracking moment values are within a plausible range.

$$M_{Rd.s.est} = 0.9 \cdot d \cdot COE \cdot f_t \cdot A_{t,l} \tag{8.17}$$

For the reference C00 concrete, the value of M_{Rd} is:

$$M_{Rd.s.est.C00} = 0.9 \cdot 15\,mm \cdot 0.33 \cdot 2500N/mm^2 \cdot 142\,mm^2 = 1.58kNm \quad (8.18)$$

whereby the value of Coefficient of Efficiency, COE, was calculated with Eq. 7.2. Then a ratio of the cracking moment and the estimated ultimate strength can be calculated and it is equal to:

$$\frac{M_{cr.s.C00}}{M_{Rd.s.est.C00}} = \frac{0.48\,kNm}{1.58\,kNm} = 30\% \quad (8.19)$$

8.4 Summary

Flexural tests were performed on beam and slab elements. In each case, four point bending tests were performed, in beams and large-scale slabs with a span of 1800 mm and mid-scale slabs with a span of 1000 mm. Three concrete mixtures were compared in all tests, reference C00 concrete, and two expansive concrete mixtures, C15 and C20. Load, deflection, strains in the reinforcement and on the concrete surface were measured with various methods and the cracking pattern was analysed with a DIC method. Experiments were carried out in several loading cycles until a planned deflection was achieved and then stopped after unloading.

In each case, it was proven that the cracking moment was increased in expansive concrete specimens. The mean increase was 14% and 66% in C15 and C20 beams, 24% and 26% in C15 and C20 mid-scale slabs and 53% and 35% in C15 and C20 large-scale slabs (here however the value for C20 slabs can be underestimated due to the limited number of specimens). Simultaneously, deflection corresponding with the load at initial crack formation was also increased. The improvement in the cracking moment suggests that chemical prestress was introduced by the restrained expansion of concrete. Taking into account the different tensile strength of both concretes, prestress was analysed and prestress level calculated for C20 beam and slab elements.

Analysis of prestress for elements subjected to flexure, i.e. beams and large slabs revealed that in both cases elements made of C20 concrete are chemically prestressed and the prestress effect is not neglectable, reaching in concrete 0.89 N/mm^2 in large slabs (Eq. 8.16) and 1.98 N/mm^2 in beams (Eq. 8.4). Nevertheless, contrary to tensile specimens, calculations performed here do not correspond well with prestress assessments made based on measured expansive strains, which gave the result of 1.01 N/mm^2 in C20 beams.

Despite the relatively low degree of reinforcement of the beams (0.44%) and slabs (0.47%) and the not particularly high level of prestress, a significant improvement in the load that can be absorbed during initial crack formation was achieved. Nonetheless, for mid-scale slabs with textile reinforcement analysis of the cracking pattern has proven that the crack spacing was also increased for expansive concrete elements when compared with the reference concrete. Furthermore, the DFOS used not only enabled monitoring of the reinforcement and concrete strains during the setting and hardening process, but also early detection of crack formation in the flexural tests. Crack locations detected with optical sensors were in perfect agreement with cracks observed later on with the DIC method or visible on the surface of the elements.

Chapter 9

Discussion and conclusions

The aim of the thesis was to experimentally assess the behaviour of expansive concrete elements with textile reinforcement. An assumption was made, that using CSA expansive admixture in various amounts can lead to introducing chemical prestress in the textile reinforced elements. To achieve this goal, expansion of free and restraint specimens was measured and analysed and concrete mechanical properties were investigated. Afterwards, pull-out, tensile and flexural tests were performed to investigate the introduced chemical prestressing.

9.1 Fulfillment of research objectives

Three objectives were formulated within the scope of this work:

Objective 1. To prove whether chemical prestressing of concrete members with carbon textile reinforcement is possible.

Expansion measurements proved, that self-compacting concrete with CSA admixture exhibits expansion during its early days during and after hydration. When the expansion is restrained by internal reinforcement, compressive stresses are introduced into test specimens, and the chemical prestressing phenomena can be confirmed on specimens with textile reinforcement.

The best way to observe this prestressing effect is to perform tensile tests on TRC specimens. In uniaxial stress state, taking into account differences in tensile strength of plain concrete, the differences between reference and expansive concrete mixtures in their cracking load capacity can be attributed directly to the prestressing effect.

In specimens prepared for tensile tests, shrinkage and expansion strains were measured during hardening and maturing, and the difference between reference, C00, and expansive, C20, specimens was equal to 0.78 mm/m, which in terms of stresses mean a prestress of concrete with a magnitude of 0.75 N/mm^2 (Eq. 6.5).

Afterwards, in tensile tests performed on the same specimens it was observed that specimens made of expansive concrete show a significantly higher cracking load than reference concrete specimens (+18% and +35% increase for C15 and C20 expansive concrete mixtures). Furthermore, optical sensors glued to the internal textile reinforcement allowed to measure textile strains during tensile tests, and these strains at the moment of the first crack also increased (by 22% for C15 and 73% for C20 specimens). Analytical assessment of the concrete prestress results in a value of $\sigma_{\text{pre.T.C20}} = 0.77$ N/mm^2 (Eq. 7.6), which corresponds well with the prestress assumed from expansion measurements (0.75 N/mm^2).

Therefore, tensile tests have proven that chemical prestressing is indeed possible, also for specimens with textile reinforcement.

Objective 2. To quantify the introduced prestressing forces and the influence of prestressing on serviceability limit state, focusing on cracking load.

The second objective was to quantify the prestress which was introduced by expansion. For this purpose, flexural tests were conducted on beam and slab specimens.

First of all, mid-scale slab specimens with carbon textile reinforcement were tested as a feasibility study of chemical prestressing. After promising results of load-deflection relationship were obtained, a study on large-scale slabs with textile reinforcement and on reference beam specimens with uniaxial steel reinforcement was conducted. Elements were tested in flexure as it is the dominant load case for thin TRC members used as façade elements. However, under flexural load the prestress is more difficult to assess, because there are additional factors that need to be considered, for example a phenomenon of increased bending crack resistivity for expansive concrete specimens (Chapter 3.4).

When beams were subjected to a four-point bending test, a substantial increase in cracking moment was observed. Beams made of expansive concrete had 14% (C15) or 66% (C20 specimens) higher cracking load in comparison to reference specimens (C00). The cracking pattern was investigated with numerous methods, including DIC, strain gauges installed on the steel rebar and distributed fibre optic sensors (DFOS) measuring strains of concrete and rebar. All of these methods provided a good agreement. Finally, concrete prestress in C20 beams was estimated as equal to $\sigma_{pre.B.C20} = 1.98$ N/mm^2 (Eq. 8.4).

For large scale slabs, flexural test have also proven that expansive concrete specimens sustain higher cracking load than reference concrete (+53% and +35% for C15 and C20 specimens). At the same time, deflection corresponding with the cracking load also increased for expansive concrete specimens. Calculated concrete prestress reached $\sigma_{pre.S.C20} = 0.89$ N/mm^2 (Eq. 8.16).

Although beams and slabs have the same length and reinforcement ratio, there is a difference between calculated concrete prestress and the ratio of $\sigma_{pre.B.C20}$ to $\sigma_{pre.S.C20}$ equals 2.22. Such a high difference might be caused by several factors. First of all, the difference in reinforcement type, either uniaxial in beams or in a grid form in slabs. As the slabs have also transversal reinforcement which acts as additional restraint, different behaviour might be expected. Secondly, the difference in geometry, as in the beams much higher concrete cover was provided over the reinforcement and the expansive strains in concrete could have been transferred better over the volume. In slabs, a very low concrete cover of 15 mm was provided. Another reason might be the difference in bond strength between concrete and the reinforcement in elements with textile reinforcement (maximum bond strength of 7.46 N/mm^2 as described in Chapter 7) in comparison with almost four times higher maximum bond strength of 29.6 N/mm^2 for specimens made of C20 concrete with steel 12mm-rebar (Zdanowicz and Hansen, 2019).

In comparison of calculated prestress of large-scale slabs and tensile elements, which both had textile reinforcement, the latter show a lower value (0.77 N/mm^2 vs 0.89 N/mm^2). This difference can be attributed to the low length of the tensile specimens when compared with large-scale slabs (450 mm vs 2000 mm), which resulted in a lower transfer length and does not allow for achieving similar expansive strains. Expansion of tensile specimens was indeed lower than the expansion of slabs (Chapter 6).

Objective 3. To analyse the influence of chemical prestressing on other aspects relevant for the design of structural members.

Chemical prestress or the use of CSA expansive admixture influences not only cracking load under tension or flexure. The last objective was to determine changes in the bond behaviour, as well as in the mechanical properties of the concrete itself and the tensile behaviour of chemically prestressed specimens.

Bond behaviour

The bond strength of specimens made of expansive concrete was superior to the reference concrete specimens (5.99 N/mm^2 for C00 series and 7.46 N/mm^2 for C20 series). A hypothesis is proposed that the higher bond strength of expansive concrete can be connected with the increase in the density of the expansive concretes. In literature, an effect of expansion is described, where the concrete becomes denser over time due to the storage of the expansion material in concrete pores. By increasing the density, a larger adhesive surface can form around the textile reinforcement. The matrix forms also fewer defects or air voids around the roving (Aroni et al., 1968; Jia et al., 2016). This could explain the higher pull-out force and bond strength associated with adhesion.

However, also prestressing of carbon roving might improve bond properties. The effect of higher bond strength due to prestressing of carbon rovings was described in (Krüger et al., 2001), who detected that prestressing led to a bundling effect as contact zones of the inner filaments increased and the geometry of single rovings was changed.

Pull-out investigations on specimens with textile reinforcement and expansive admixtures also confirmed that expansion improves the bond behaviour and increases the ultimate pull-out force (Wang et al., 2016).

Therefore, based on literature analysis and own experimental work, it becomes clear that expansive concrete will achieve at least the same or presumably even greater bond strength than the reference concrete without the addition of an expansive admixture.

Concrete properties

The addition of expansive admixture influenced also the mechanical properties of concrete. Compressive strength was increased for both mixtures applied, C15 and C20 (+7.1% and +12.0% in reference to C00), however when a specific threshold was achieved, here defined as 25% of expansive admixture, a decrease of -17% of the compressive strength was observed. On the other hand, the modulus of elasticity was almost not influenced by the addition of expansive admixture (-1.1% and -3.4% for C15 and C20 speciemens). The tensile flexural strength of plain concrete was also superior for C20 specimens when compared with C00 (+18% increase).

Strain residuum phenomenon

In tensile specimens with a textile reinforcement, besides the visible improvement in cracking load, a so-called strain residuum phenomenon was observed. It means that after two loading and unloading cycles even in the theoretically elastic state of a linear stress-strain relationship, some deformation remain in the expansive concrete specimens. This strain residuum becomes higher with increasing expansive admixture amount.

9.2 Overall performance index

Several areas of structural behaviour were analysed for reference and chemically prestressed concrete specimens. To be able to compare the results for C00, C15 and C20 series, a so-called overall performance index (PI) will be introduced. This index is understood as a ratio of areas of tetragons created by values achieved in various experimental tests (cracking load for slabs, beams and tensile specimens and bond strength) for each series (Eq. 9.1 and 9.2). Figure 9.1 shows a graphical representation of the PI. Note that as for C15 no data are available for bond strength, a mean value of the C00 and C20 series is assumed.

$$PI_{C20/C00} = \frac{A_{C20}}{A_{C00}} = \frac{231.39}{114.06} = 2.03 \tag{9.1}$$

$$PI_{C15/C00} = \frac{A_{C15}}{A_{C00}} = \frac{157.75}{114.06} = 1.38 \tag{9.2}$$

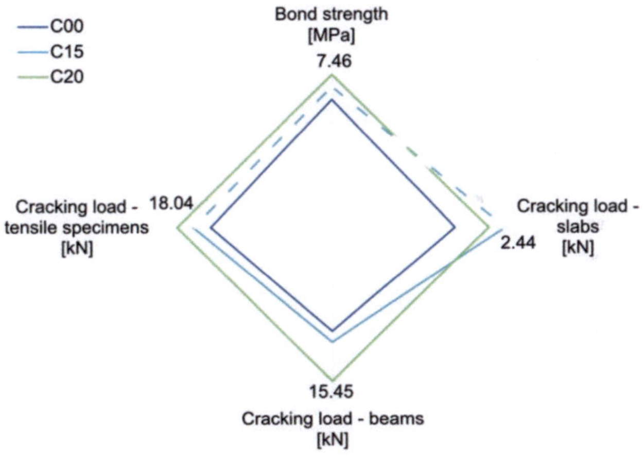

Figure 9.1: Overall performance index of C00 and C20 series

In the case of the results obtained within this thesis, the Performance Index is equal to 2.03 for C20 and 1.38 for C15. C15 specimens performed well, although the difference here was not significantly better. For C20 series it can be interpreted that specimens performed in structural tests in total two times better than reference specimens. Even taken into account differences in concrete mechanical properties (higher compressive and tensile strength and lower modulus of elasticity), the PI value shows an advantage in using chemically prestressed specimens.

In structural tests specimens made of expansive concrete behave better than reference concrete specimens. Although there are differences between specific types of tests and various factors influence their different behaviour under tension or flexure, Figure 9.2 shows that an increase in the performance of chemically prestressed specimens is visible.

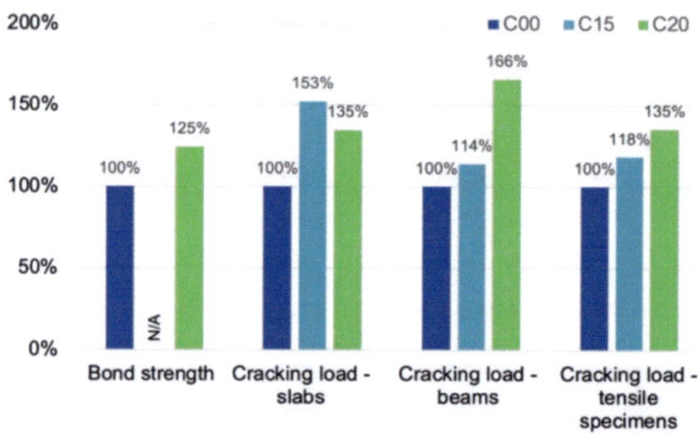

Figure 9.2: Comparison of C00, C15 and C20 series

9.3 Outlook

Although this thesis takes a comprehensive approach to analyse the structural behaviour of chemically prestressed elements on various levels, the complexity of analysed phenomena enforced multiple assumptions, models and simplifications. Also, lacking investigations and issues which were not addressed within this thesis work, described in the Introduction, limit the possibility to apply these findings in the industrial applications at this stage. Therefore, the following issues are recommended for future work:

- Calculations of prestress based on comparison of test results for both series should be validated with analytical models.
- Further research, focused on the long-term stability and durability of chemically prestressed specimens is necessary.
- Creep and shrinkage tests are needed which allow to assess the prestress losses.
- A possible improvement in the magnitude of prestress should be investigated, for instance with consideration of longer water curing or

alternative methods for supplying water (prolonged curing in water or moisture curing was not realized, as it is difficult to realize in the industrial applications).

- The extent of reaction and re-expansion possibility should be checked for the mixtures and geometries tested.
- Also, the influence of environmental conditions on chemically prestressed elements is necessary for successful field applications.
- Finally, analytical models which will allow quantifying estimated prestress basing on simple expansion measurements are compulsory.

Although there is still plenty of open research questions and gaps before chemical prestressing could be effectively applied, the work presented in this thesis confirmed that further developement of this technology might be beneficial for numerous applications of elements with carbon textile reinforcement. Prestressing of textile reinforcement might be performed in a novel way, using a technology that has not found its field of application before, and it can be beneficial for structural performance and usability.

Appendix A

Large slab specimens - additional results

The appendix contains additional results for large-scale slabs, not presented in Chapter 8.3.

Load-deflection diagrams for slabs

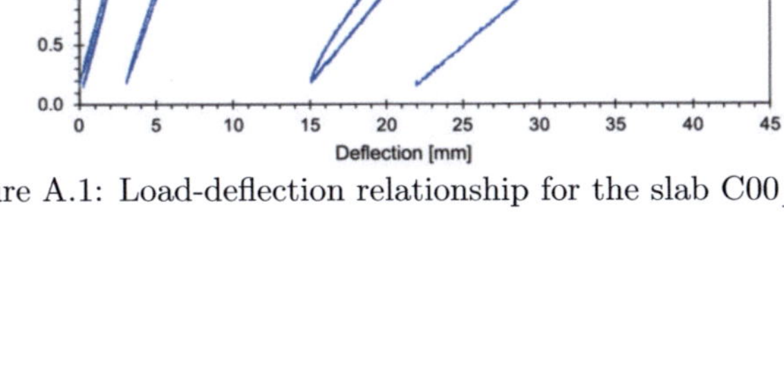

Figure A.1: Load-deflection relationship for the slab C00_S_a

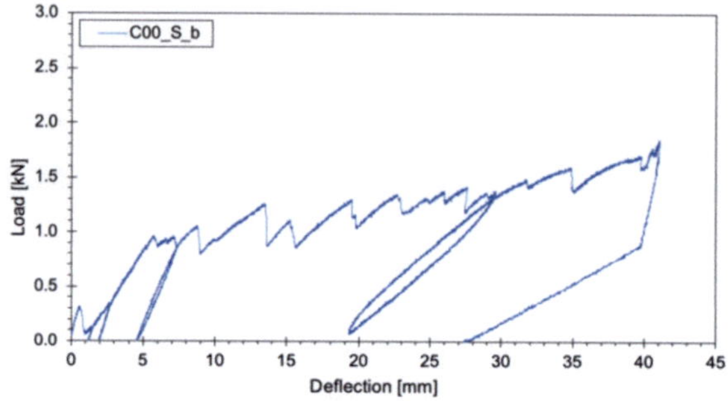

Figure A.2: Load-deflection relationship for the slab C00_S_b

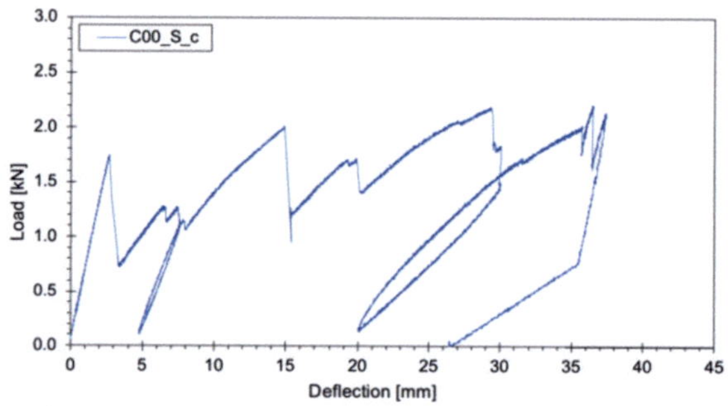

Figure A.3: Load-deflection relationship for the slab C00_S_c

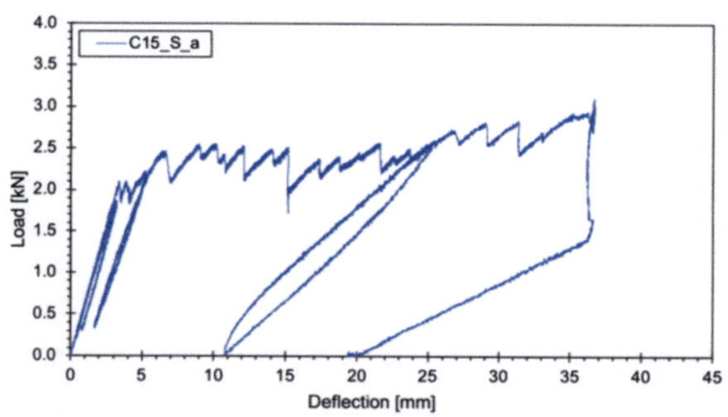

Figure A.4: Load-deflection relationship for the slab C15_S_a

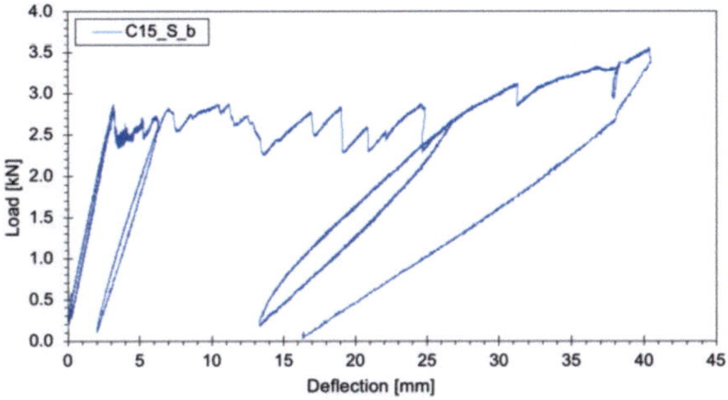

Figure A.5: Load-deflection relationship for the slab C15_S_b

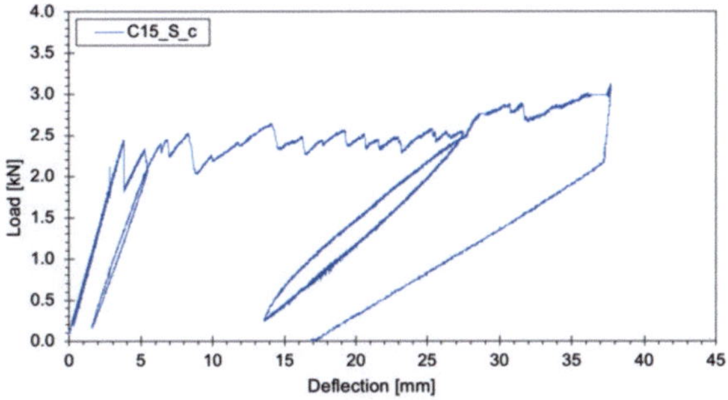

Figure A.6: Load-deflection relationship for the slab C15_S_c

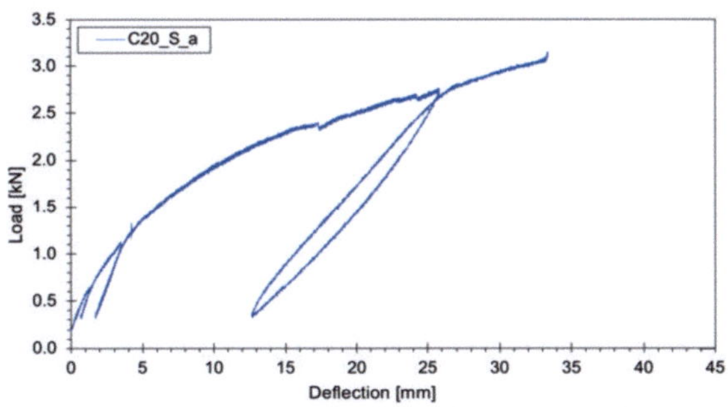

Figure A.7: Load-deflection relationship for the slab C20_S_a

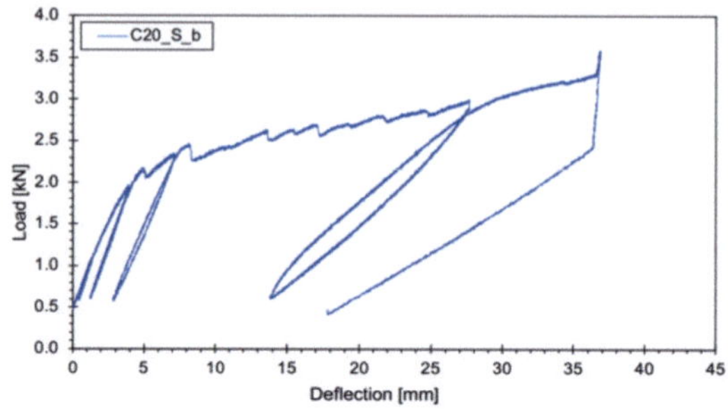

Figure A.8: Load-deflection relationship for the slab C20_S_c

DIC results for large slabs

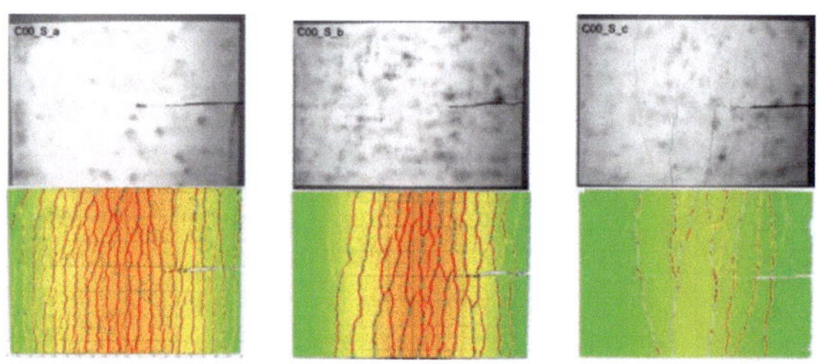

Figure A.9: DIC results - cracking pattern for slabs from C00_S series

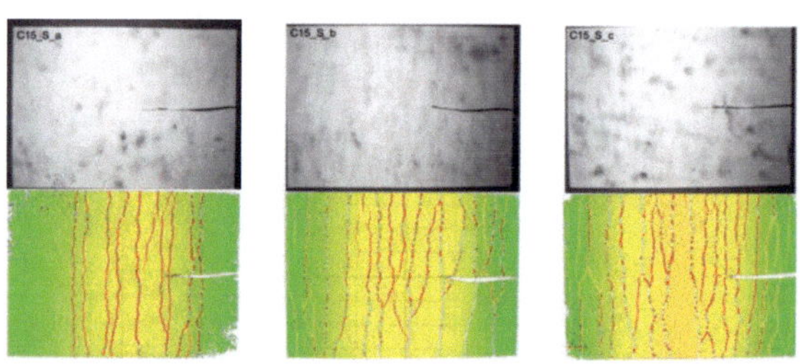

Figure A.10: DIC results - cracking pattern for slabs from C15_S series

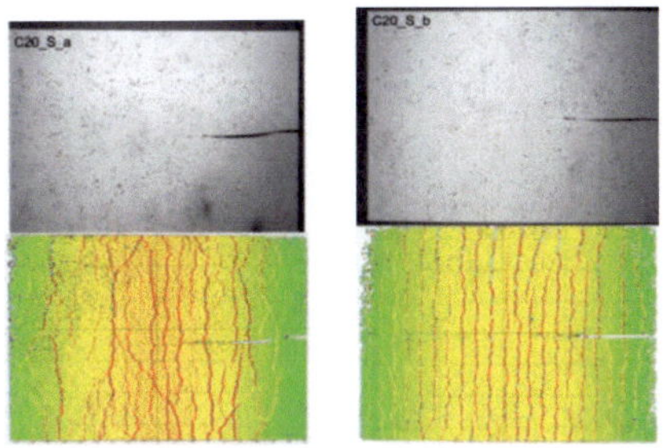

Figure A.11: DIC results - cracking pattern for slabs from C20_S series

DFOS results for large slabs

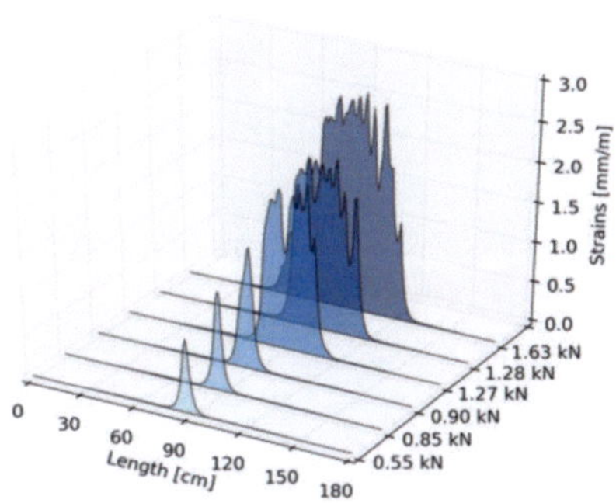

Figure A.12: DFOS measurements - cracking pattern for the slab C00_S_b

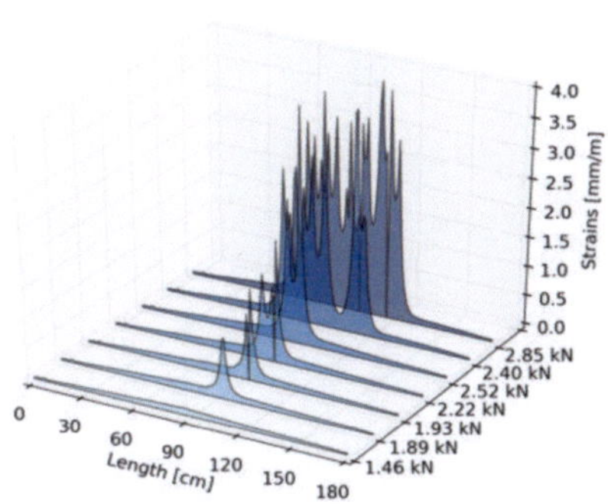

Figure A.13: DFOS measurements - cracking pattern for the slab C15_S_a

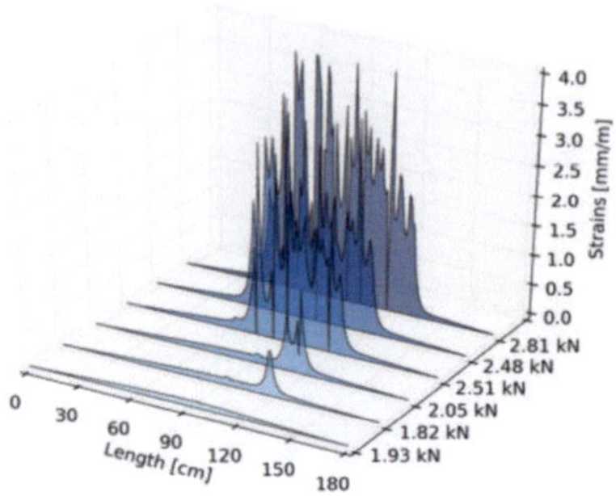

Figure A.14: DFOS measurements - cracking pattern for the slab C15_S_c

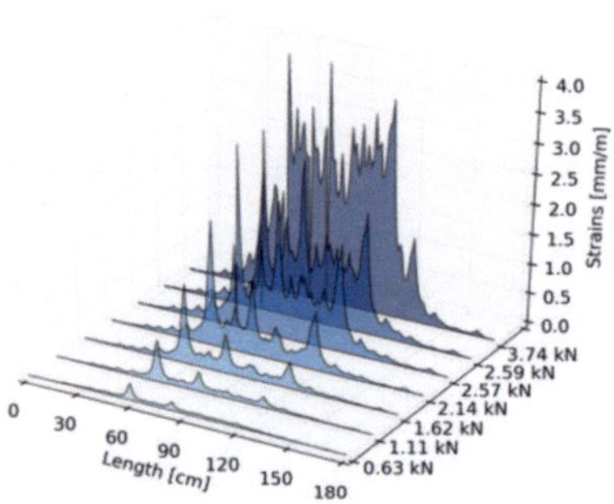

Figure A.15: DFOS measurements - cracking pattern for the slab C20_S_a

DFOS and DIC correlation of large slabs

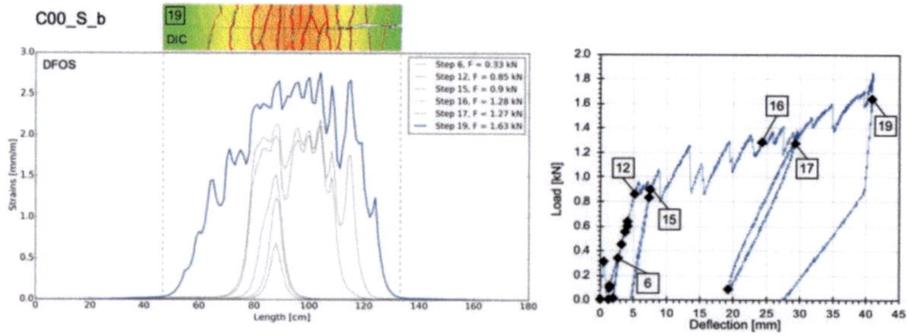

Figure A.16: Comparison of measurements results from DFOS sensors with DIC method for the slab C00_S_b, load step xx

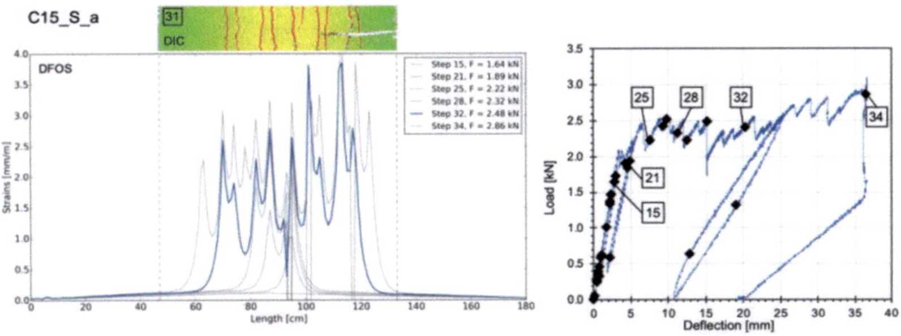

Figure A.17: Comparison of measurements results from DFOS sensors with DIC method for the slab C15_S_a, load step xx

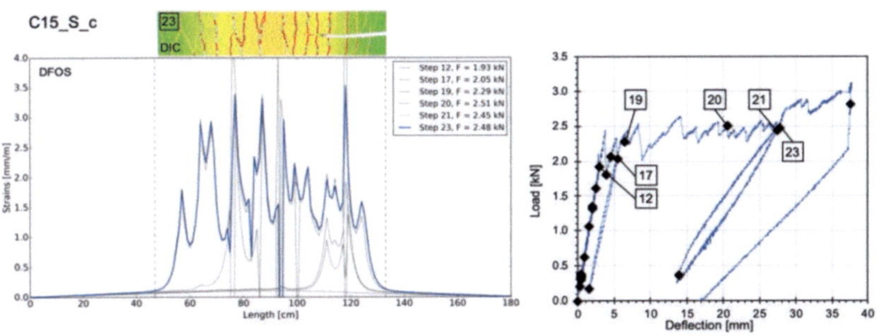

Figure A.18: Comparison of measurements results from DFOS sensors with DIC method for the slab C15_S_c, load step xx

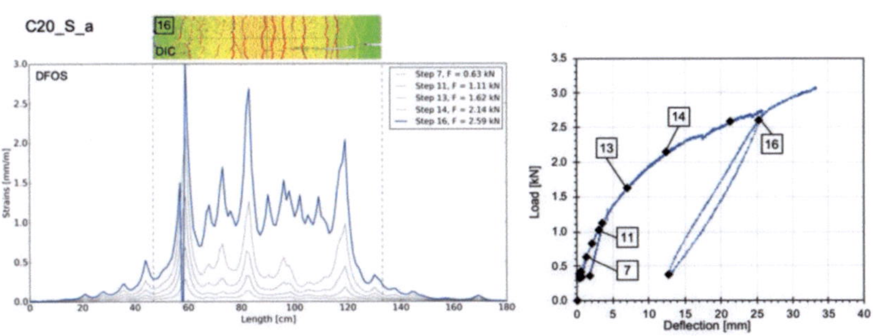

Figure A.19: Comparison of measurements results from DFOS sensors with DIC method for the slab C20_S_a, load step xx

Appendix B

Beam specimens - additional results

The appendix contains additional results for beam specimens, not presented in Chapter 8.1.

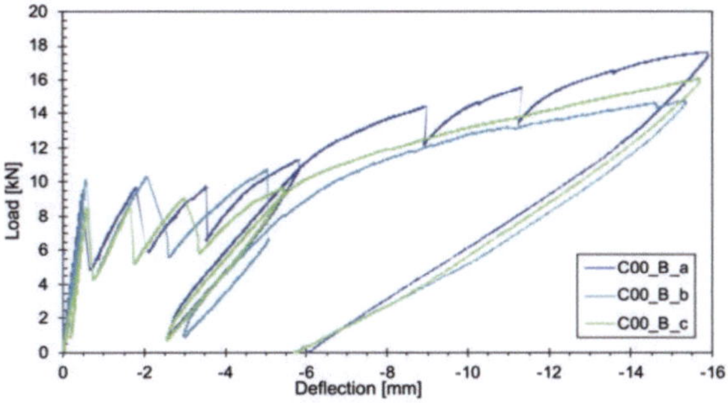

Figure B.1: Load-deflection diagram for specimens from the series C00

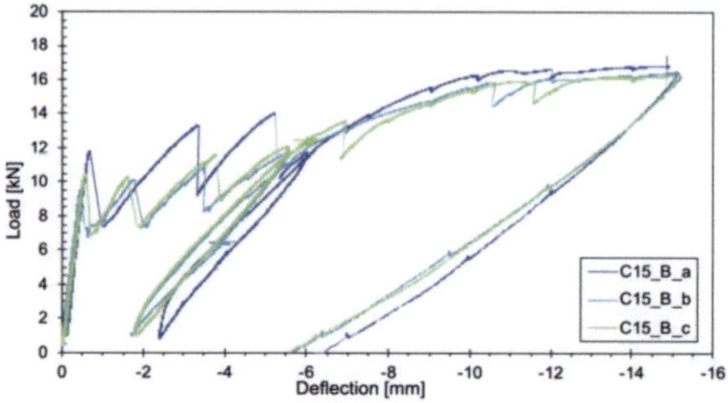

Figure B.2: Load-deflection diagram for specimens from the series C15

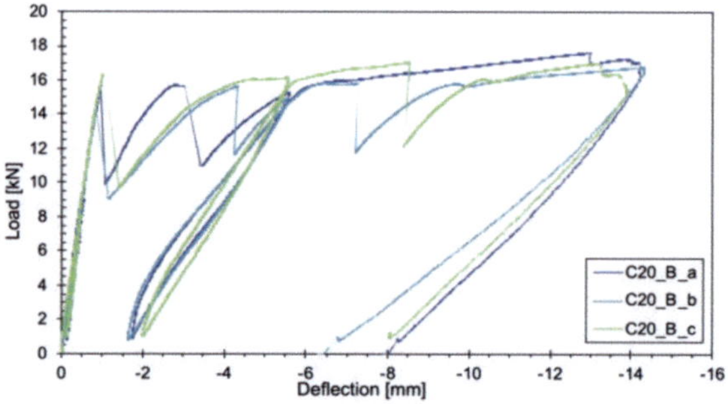

Figure B.3: Load-deflection diagram for specimens from the series C20

Appendix C

Tensile specimens - additional results

The appendix contains additional results for tensile specimens, not presented in Chapter 7.2.

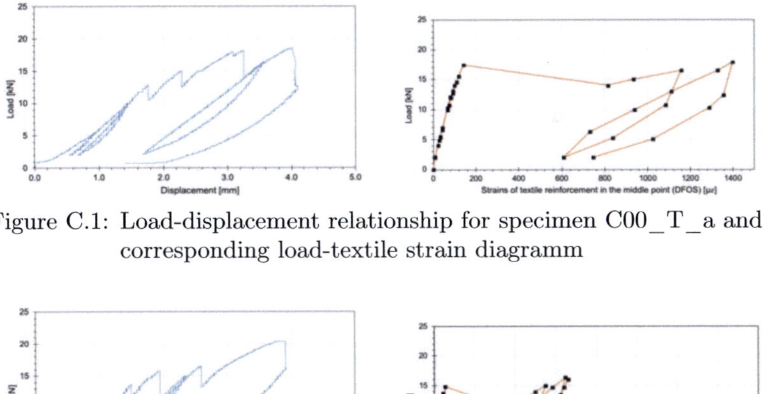

Figure C.1: Load-displacement relationship for specimen C00_T_a and corresponding load-textile strain diagramm

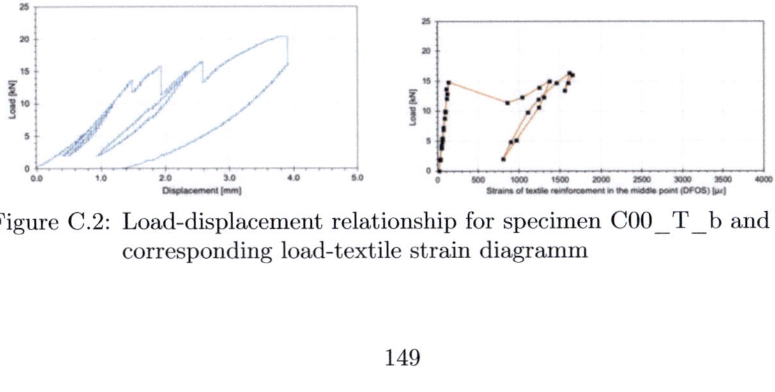

Figure C.2: Load-displacement relationship for specimen C00_T_b and corresponding load-textile strain diagramm

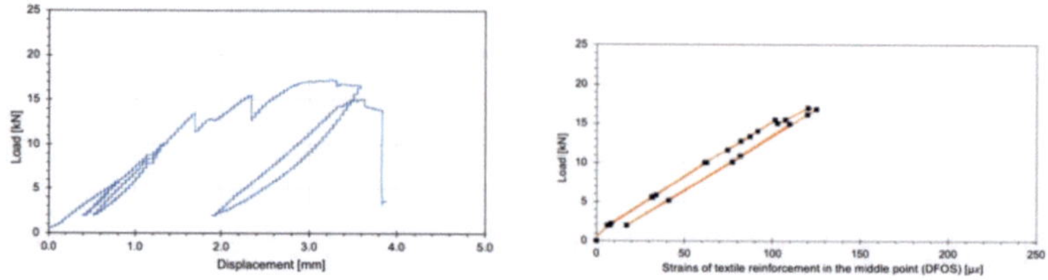

Figure C.3: Load-displacement relationship for specimen C15_T_b and corresponding load-textile strain diagramm

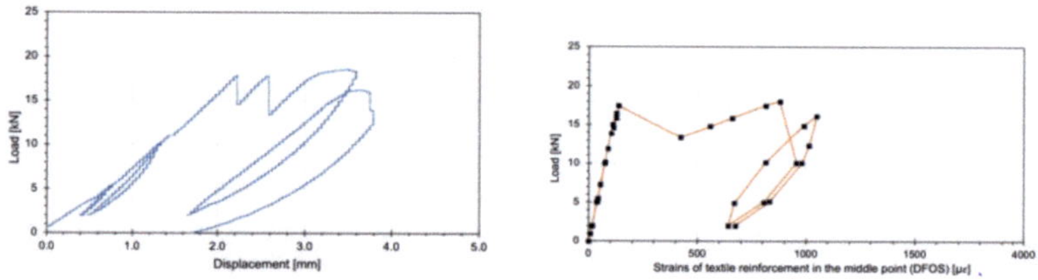

Figure C.4: Load-displacement relationship for specimen C15_T_c and corresponding load-textile strain diagramm

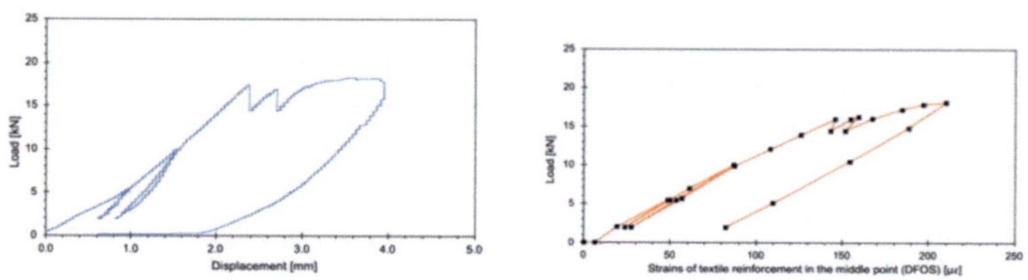

Figure C.5: Load-displacement relationship for specimen C20_T_b and corresponding load-textile strain diagramm

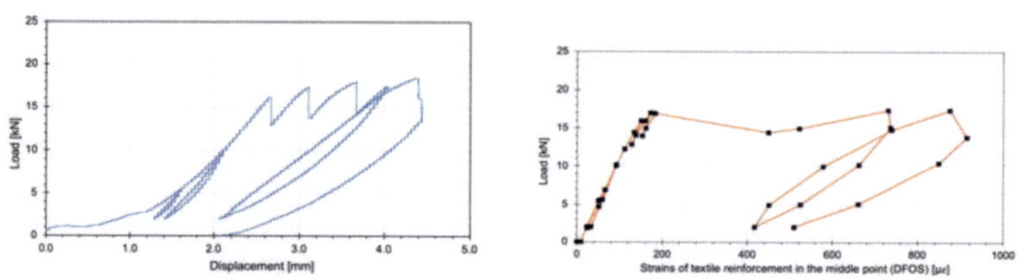

Figure C.6: Load-displacement relationship for specimen C20_T_c and corresponding load-textile strain diagramm

List of Figures

List of Tables

Bibliography

ACI Committee 223 (1970), 'Expansive Cement Concretes - Present State of Knowledge', *ACI Structural Journal* 67, 583–610.

American Concrete Institute (2010), *ACI 223R-10 Guide for the use of shrinkage-compensating concrete*, Vol. 1, American Concrete Institute.

Aroni, S., Bertero, V. and Polivka, M. (1968), 'Chemically Prestressed Concrete', *PCI Journal* 13(5), 22–35.

ASTM (2016), 'ASTM C403 / C403M - 16: Standard Test Method for Time of Setting of Concrete Mixtures by Penetration Resistance'.

ASTM C878 (2014), 'Standard Test Method for Restrained Expansion of Shrinkage-Compensating Concrete'.

Bassil, A., Chapeleau, X., Leduc, D. and Abraham, O. (2020), 'Concrete crack monitoring using a novel strain transfer model for distributed fiber optics sensors', *Sensors* 20(8), 1–17.

Battaglia, I. (2012), Experimental Use of Type K Cement Concrete in Wisconsin Highway Bridge Decks, Report No. FEP-04-12, Technical report, Wisconsin Department of Transportation.

Benuska, K. L., Bertero, V. V. and Polivka, M. (1971), 'Self-stressed concrete for precast building units', *PCI Journal* 16(2), 72–84.

Bergmeister, K. (1999), 'Vorspannung von Kabeln und Lamellen aus Kohlenstoffasern', *Beton- und Stahlbetonbau* 94, 20–26.

Bertero, V. (1967), 'Curing Effects on Expansion and Mechanical Behavior of Expansive Cement Concrete', *Journal of the American Concrete Institute* 64(2), 84–96.

Billig, K. (1952), *Prestressed Concrete*, Macmillan & Co., London.

Bischoff, D. and Toepel, A. (2004), Laboratory Testing of Portland Cement Concrete Patch Material, Modified To Reduce or Eliminate Shrinkage, Technical report, WisDOT Research Study WI-02-05.

Bizzozero, J., Gosselin, C. and Scrivener, K. L. (2014), 'Expansion mechanisms in calcium aluminate and sulfoaluminate systems with calcium sulfate', *Cement and Concrete Research* 56, 190–202.

Breitenbücher, R., Ebensperger, L. and Springenschmid, R. (1992), 'Wirksamkeit von Quellzusätzen im Beton für den Ausgleich von Zwangspannungen', *Beton- und Stahlbetonbau* 87(9), 217–223.

Bunke, N. (1991), 'Prüfung von Beton, Empfehlung und Hinweise als Ergänzung zu DIN 1048 (DAfStb Heft 422)'.

Carballosa, P., García Calvo, J. L., Revuelta, D., Sánchez, J. J. and Gutiérrez, J. P. (2015), 'Influence of cement and expansive additive types in the performance of self-stressing and self-compacting concretes for structural elements', *Construction and Building Materials* 93, 223–229.

Champenois, J. B., Dhoury, M., Cau Dit Coumes, C., Mercier, C., Revel, B., Le Bescop, P. and Damidot, D. (2015), 'Influence of sodium borate on the early age hydration of calcium sulfoaluminate cement', *Cement and Concrete Research* 70, 83–93.

Chaunsali, P. and Mondal, P. (2015*a*), 'Influence of Calcium Sulfoaluminate (CSA) Cement Content on Expansion and Hydration Behavior of Various Ordinary Portland Cement-CSA Blends', *Journal of the American Ceramic Society* 98(8), 2617–2624.

Chaunsali, P. and Mondal, P. (2015*b*), 'Influence of mineral admixtures on early-age behavior of calcium sulfoaluminate cement', *ACI Materials Journal* 112(1), 59–68.

Chaunsali, P. and Mondal, P. (2016), 'Physico-chemical interaction between mineral admixtures and OPC-calcium sulfoaluminate (CSA) cements and its influence on early-age expansion', *Cement and Concrete Research* 80, 10–20.

Cohen, M. D. (1983), 'Modeling of expansive cements', *Cement and Concrete Research* 13(4), 519–528.

Destrée, X., Cepuritis, R., Fischer, G. and Suta, M. (2018), Effect of Intrinsically Post-tensioned Steel Fibre Reinforced Concrete on the Structural Response and Design of Slabs on Grade and Suspended Slabs – from Design Stage to Application, *in* 'Proceedings of 5th International fib Congress "Smarter - Better - Stronger"', Melbourne, Australia.

EN 12390-1 (2012), 'EN 12390-1. Testing hardened concrete. Shape, dimensions and other requirements for specimens and moulds'.

EN 12390-13 (2013), 'EN 12390-13. Testing hardened concrete. Determination of secant modulus of elasticity in compression'.

EN 12390-3 (2019), 'EN 12390-3:2019. Testing hardened concrete. Compressive strength of test specimens'.

EN 206:2013 (2013), 'EN 206:2013. Concrete. Specification, performance, production and conformity.'.

fib Fédération internationale du béton (2013), *fib Model Code for Concrete Structures 2010*, Ernst & Sohn.

Fischer, O., Thoma, S. and Crepaz, S. (2019), 'Quasikontinuierliche faseroptische Dehnungsmessung zur Rissdetektion in Betonkonstruktionen', *Beton- und Stahlbetonbau* 114, 1–10.

Folliard, K. J., Ohta, M., Rathje, E. and Collins, P. (1994), 'Influence of mineral admixtures on expansive cement mortars', *Cement and Concrete Research* 24(3), 424–432.

Garcia Calvo, J. L., Revuelta, D., Carballosa, P. and Gutierrez, J. P. (2017), 'Comparison between the performance of expansive SCC and expansive conventional concretes in different expansion and curing conditions', *Construction and Building Materials* 136, 277–285.

Habel, W. R. and Bismarck, A. (2000), 'Optimization of the adhesion of fiber-optic strain sensors embedded in cement matrices; a study into long-term fiber strength', *Journal of Structural Control* 7(1), 51–76.

Hanein, T., Galvez-Martos, J. L. and Bannerman, M. N. (2018), 'Carbon footprint of calcium sulfoaluminate clinker production', *Journal of Cleaner Production* 172, 2278–2287.

Hargis, C. W., Lothenbach, B., Müller, C. J. and Winnefeld, F. (2018), 'Further insights into calcium sulfoaluminate cement expansion', *Advances in Cement Research* pp. 1–18.

He, H. N., Dong, W. and Wu, Z. M. (2010), 'Study on Long-Term Expansive Deformation of Self-Stressing Concrete with Combined Restrictions of Steel Fibers and Steel Bar', *Key Engineering Materials* 452, 533–536.

He, H. N., Wang, B. X. and Lin, J. T. (2009), 'Performance on Steel Fiber Reinforced Self-Stressing Concrete', *Key Engineering Materials* 400-402, 427–432.

Hoff, G. C. (1972), Expansive Cements and Their Use, Technical report, US Army Waterways Experiment Station, Vicksburg, Mississippi.

Hosoda, A. and Kishi, T. (2001), 'Mechanism of nonlinear behavior and crack resistancc of expansive mortar', *Translation from Proceedings of JSCE* 52(683), 219–235.

Hosoda, A., Minoru, M., Tanimura, M., Kanda, T., Sakai, E. and Kishi, T. (2011), Technical committee on perfromance evaluation of high performance expanisve concrete and system for crack control, Technical report.

Hosoda, A., Morioka, M., Tanimura, M., Kanda, T., Sakai, E. and Kishi, T. (2011), 'Technical Committee on Performance Evaluation of High Performance Expansive Concrete and System for Crack Control'.

Hu, Y., Li, W., Ma, S. and Shen, X. (2017), 'Influence of borax and citric acid on the hydration of calcium sulfoaluminate cement', *Chemical Papers* 71(10), 1909–1919.

Iijima, H., Sakai, Y. and Kishi, T. (2013), 'Mechanical performance of expansive concerte in uniaxial tension test with different cement types, ages and curing conditions', *Thirteenth East Asia-Pacific Conference on Structural Engineering and Construction (EASEC-13)* .

Ishikawa, Y. and Tanabe, T. (2010), Theoretical development of CP method in predicting expansive cement concrete cracking, *in* 'Proceedings of FraMCoS-7: Fracture Mechanics of Concrete and Concrete Structures'.

Ito, H., Maruyama, I., Tanimura, M. and Sato, R. (2004), 'Early age deformation and resultant induced stress in expansive high strength concrete', *Journal of Advanced Concrete* 2(June), 155–174.

Japan Society of Civil Engineering (1994), 'Recommended Practice for Expansive Concrete'.

Jia, L.-l., Wu, J., Zhu, J.-l., Yang, L., Jawaid, M. and Kenawy, E.-R. (2016), 'Preparation and Expansion Properties Analysis of C60 Expansive Self-compacting Concrete', *MATEC Web of Conferences* 67, 7032.

JIS (1999), 'JIS A 6202 "Expasive Additive for Concrete. Method of Testing Restrained Expansion and Shrinkage of Expansive Concrete"'.

José Oliveira, M., Ribeiro, A. B. and Branco, F. G. (2014), 'Combined effect of expansive and shrinkage reducing admixtures to control autogenous shrinkage in self-compacting concrete', *Construction and Building Materials* 52, 267–275.

Justnes, H. (2007), Principles of making cement with reduced CO2 emission: State of the art. Sintef report No. SBF BK A07019, Technical report, SINTEF Building and Infrastructure, COIN – Concrete Innovation Centre.

Keeton, J. R. (1973), Expansive cement concretes for naval construction, Report TN-1264, Technical Report March, Naval Civil Engineering Laboratory, Port Hueneme, California.

Keil, F. (1971), *Zement. Herstellung und Eigenschaften.*, Vol. 53, Springer Verlag, Berlin, Heidelberg, New York.

Keith, F., Bentley, C., Walker, W. and Holland, J. (2006), 'Shrinkage-Compensating Concrete Pavements Perform Well', *Concrete International* 28(01), 47–51.

Kishi, T., Hosoda, A., Gurung, C. B. and Kittiwuttichusinp, S. (2000), 'Effect of Drying/Autogenous Shrinkage on Ductility/Fracture Mode of Beam and Self-Repairing Function of Expansive Agent', *Japan Concrete Institute* 22(3), 511–516.

Klein, A., Karby, T. and Polivka, M. (1961), 'Properties of an Expansive Cement for Chemical Prestressing', *Journal of the American Concrete Institute* 58(7), 59–82.

Krenkler, K. (1980), *Chemie des Bauwesens*, Springer Verlag.

Krüger, M. (2004), Vorgespannter textilbewehrter beton, Dissertation, Universität Stuttgart, Stuttgart.

Krüger, M., Reinhardt, H. W. and Fichtlscherer, M. (2001), 'Bond behaviour of textile reinforcement in reinforced and prestressed concrete', *Otto-Graf-Journal* 12, 33–50.

Kulas, C. (2013), Zum Tragverhalten getränkter textiler Bewehrungselemente für Betonbauteile, Dissertation, RWTH Aachen, Aachen.

Kurdowski, W. (2014), *Cement and Concrete Chemistry*, Springer Science & Business.

Kurdowski, W. and Thiel, A. (1981), 'On the role of free calcium oxide in expansive cements', *Cement and Concrete Research* 11(1), 29–40.

Leemann, A. A. and Lura, P. (2014), Creep and shrinkage of SCC, *in* K. H. Kayat and G. De Schutter, eds, 'Mechanical Properties of Self-Compacting Concrete, STAR of the RILEM TC 228-MPS', RILEM Publications, pp. 1–28.

Li, S. T. (1965), 'Expansive Cement Concrete ...a new idea with exciting possibilities', *The Aberdeen Group Publications* No. C650207, 1–3.

Lin, T. Y. and Klein, A. (1963), 'Chemical Prestressing of Concrete Elements Using Expanding Cements', *Journal of the American Concrete Institute* 60(9), 1187–1218.

Lobo, C. and Cohen, M. D. (1992), 'Effects of Silica Fume on Expansion Characteristics of Expansive Cement Pastes', *ACI Materials Journal* 89(5), 481–490.

Lorenz, E. (2014), Endverankerung und übergreifung textiler bewehrungen in betonmatrices., Dissertation, Technische Universität Dresden.

Lorenz, E. and Ortlepp, R. (2011), Bond Behavior of Textile Reinforcements - Development of a Pull-Out Test and Modeling of the Respective Bond versus Slip Relation, Springer.

Loser, R. and Leemann, A. (2009), 'Shrinkage and restrained shrinkage cracking of self-compacting concrete compared to conventionally vibrated concrete', *Materials and Structures* 42(1), 71–82.

Lothenbach, B., Le Saout, G., Gallucci, E. and Scrivener, K. (2008), 'Influence of limestone on the hydration of Portland cements', *Cement and Concrete Research* 38(6), 848–860.

Luna Innovations Inc. (2013), 'User Guide: Optical Backscatter Reflectometer 4600'.

Maekawa, K., Pimanmas, A. and Okamura, H. (2003), *Non-linear mechanics of reinforced concrete*, Spon Press, Taylor & Francis Group.

Magnel, G. (1956), *Theorie und Praxis des Spannbetons*, Bauverlag GmbH, Berlin.

Mailvaganam, N. P. (1995), Miscellaneous Admixtures, *in* V. S. Ramachandran, ed., 'Concrete admixtures handbook', Building materials science series, Noyes Publications, Park Ridge, N.J., U.S.A.

Martin, L. H. J., Winnefeld, F., Müller, C. J. and Lothenbach, B. (2015), 'Contribution of limestone to the hydration of calcium sulfoaluminate cement', *Cement and Concrete Composites* 62, 204–211.

Mather, B. (1984), 'A discussion of the paper "Theories of expansion in sulfoaluminate-type expansive cements: schools of thought" by M.D. Cohen', *Cement and Concrete Research* 14, 603–609.

Mehta, P. K. (1967), Expansion Characteristics of Calcium Sulfoaluminate Hydrates, *in* '68th Annual Meeting of The American Ceramic Society', pp. 204–208.

Mehta, P. K. (1973), 'Mechanism of expansion associated with ettringite formation', *Cement and Concrete Research* 3(1), 1–6.

Möll, H. (1954), *Spannbeton*, Berliner Union.

Mufti, A. A. (2002), 'Structural Health Monitoring of Innovative Canadian Civil Engineering Structures', *Structural Health Monitoring* 1(1), 89–103.

Naaman, A. E. (2000), Self-stressing composites (Chapter 10.6), *in* 'Ferrocement and Laminated Cementitious Composites', Techno Press 3000, USA, p. 367.

Nagataki, S. (1980), 'Expansive Cement Concretes in Japan', *ACI Special Publication* 64, 43–80.

Nagataki, S. and Gomi, H. (1998), 'Expansive admixtures (mainly ettringite)', *Cement and Concrete Composites* 20(2-3), 163–170.

Nocuń-Wczelik, W., Konik, Z. and Stok, A. (2011), 'Blended systems with calcium aluminate and calcium sulphate expansive additives', *Construction and Building Materials* 25(2), 939–943.

Okada, K., Ohta, M., Nagafuchi, T., Yata, A. and Tamai, S. (1983), 'Characteristics of Expansive Concrete under Bi-Axial Restraint', pp. 182–187.

Okamura, H., Tsuji, Y. and Maruyama, K. (1977), 'Application of Expansive Concrete in Structural Elements', *Journal of the Faculty of Engineering, University of Tokyo, Series B* 34(3), 481–507.

Olek, J. and Cohen, M. D. (1991), 'Procedure for Mixture Proportioning of Type-K Expansive Cement Blends for Use in Shrinkage-Compensating Mortars', *ACI Materials Journal* 88(5), 536–543.

Peled, A., Mobasher, B. and Bentur, A. (2017), *Textile reinforced concrete*, CRC Press, Taylor {&} Francis Group.

Péra, J. and Ambroise, J. (2004), 'New applications of calcium sulfoaluminate cement', *Cement and Concrete Research* 34(4), 671–676.

Pimraksa, K. and Chindaprasirt, P. (2018), *Sulfoaluminate cement-based concrete*, Elsevier Ltd.

Polivka, M. (1973), 'Factors Influencing Expansion of Expansive Cement Concretes', *ACI Special Publication* 38, 239–250.

Reinhardt, H. W., Krüger, M. and Große, C. U. (2003), 'Concrete Prestressed with Textile Fabric', *Journal of Advanced Concrete Technology* 1(3), 231–239.

Rice, E. K. (2003), 'Restraining expansive mortar with non-metallic fibers'.

RILEM TC 232-TDT (2016), 'Recommendation of RILEM TC 232-TDT: test methods and design of textile reinforced concrete: Uniaxial tensile test: test method to determine the load bearing behavior of tensile specimens made of textile reinforced concrete', *Materials and Structures/Materiaux et Constructions* 49(12), 4923–4927.

Sahamitmongkol, R. and Kishi, T. (2011), 'Tension stiffening effect and bonding characteristics of chemically prestressed concrete under tension', *Materials and Structures* 44(2), 455–474.

Saidi, M. and Gabor, A. (2020), 'Experimental analysis of the tensile behaviour of textile reinforced cementitious matrix composites using distributed fibre optic sensing (DFOS) technology', *Construction and Building Materials* 230, 117027.

Sakai, Y. and Kishi, T. (2012), 'Mechanism Clarification and Knowledge Formation of Unique Behavior of Chemically Prestressed Concrete for Efficient Use of Expansive Concrete', *Society for Social Management Systems Internet Journal* 05.

Schmidt, J. W., Bennitz, A., Täljsten, B., Goltermann, P. and Pedersen, H. (2012), 'Mechanical anchorage of FRP tendons – A literature review', *Construction and Building Materials* 32, 110–121.

Schmidt-Thrö, G., Scheufler, W. and Fischer, O. (2016), 'Kontinuierliche faseroptische Dehnungsmessung im Stahlbetonbau', *Beton- und Stahlbetonbau* 111(8), 496–504.

Sienko, R., Bednarski, L. and Howiacki, T. (2019), 'About Distributed Internal and Surface Strain Measurements within Prestressed Concrete Truck Scale Platforms', *IOP Conference Series: Materials Science and Engineering* 471(5), 1–10.

Sieńko, R., Zych, M., Bednarski, Ł. and Howiacki, T. (2018), 'Strain and crack analysis within concrete members using distributed fibre optic sensors', *Structural Health Monitoring* 18(6), 1510–1526.

Sirtoli, D., Wyrzykowski, M., Riva, P., Tortelli, S., Marchi, M. and Lura, P. (2019), 'Shrinkage and creep of high-performance concrete based on calcium sulfoaluminate cement', *Cement and Concrete Composites* 98(2), 61–73.

Sisomphon, K., Copuroglu, O. and Koenders, E. A. B. (2012), 'Self-healing of surface cracks in mortars with expansive additive and crystalline additive', *Cement and Concrete Composites* 34(4), 566–574.

Slowik, V., Schlattner, E. and Klink, T. (2004), 'Experimental investigation into early age shrinkage of cement paste by using fibre Bragg gratings', *Cement and Concrete Composites* 26(5), 473–479.

Solidian GmbH (2017), 'Technical Datasheet solidian GRID Q142/142-CCE-38'.

Sousa Ribeiro, M. S. (1998), 'Expansive cement blend for use in shrinkage-compensating mortars', *Materials and Structures* 31(1), 400–404.

Telesca, A., Marroccoli, M., Pace, M. L., Tomasulo, M., Valenti, G. L. and Monteiro, P. J. (2014), 'A hydration study of various calcium sulfoaluminate cements', *Cement and Concrete Composites* 53, 224–232.

Terrasi, G. P. (1998), Mit Kohlenstoffasern vorgespannte Schleuderbetonrohre, Dissertation, ETH Zurich, Zurich.

Trauchessec, R., Mechling, J. M., Lecomte, A., Roux, A. and Le Rolland, B. (2015), 'Hydration of ordinary Portland cement and calcium sulfoaluminate cement blends', *Cement and Concrete Composites* 56, 106–114.

Triantafillou, T. C. (2016), *Textile Fibre Composites in Civil Engineering*, Woodhead Publishing.

Troli, R. and Collepardi, M. (2011), Shrinkage-compensating concretes for special structures, *in* '4th Intern. Conf. on Non-Traditional Cement & Concrete', pp. 27–30.

UNI (2017), UNI 8148:2017 Agenti espansivi non metallici per impasti cementizi - Determinazione dell'espansione contrastata del calcestruzzo, Technical report.

Voss, S. (2008), Ingenieurmodelle zum Tragverhalten von textilbewehrtem Beton, Dissertation, RWTH Aachen University, Aachen.

Wang, B., Jin, H., Man, T. and Wang, Q. (2016), 'Study on the mechanical property of textile reinforced self-stressing concrete sheets', *Construction and Building Materials* 107, 1–10.

Williams Portal, N. (2015), Usability of Textile Reinforced Concrete: Structural Performance, Durability and Sustainability, Dissertation, Chalmers University of Technology, Gothenburg, Sweden.

Winnefeld, F. and Lothenbach, B. (2010), 'Hydration of calcium sulfoaluminate cements - Experimental findings and thermodynamic modelling', *Cement and Concrete Research* 40(8), 1239–1247.

Winnefeld, F., Martin, L. H., Müller, C. J. and Lothenbach, B. (2017), 'Using gypsum to control hydration kinetics of CSA cements', *Construction and Building Materials* 155, 154–163.

Wyrzykowski, M., Terrasi, G. and Lura, P. (2018), 'Expansive high-performance concrete for chemical-prestress applications', *Cement and Concrete Research* 107, 275–283.

Yazdizadeh, Z., Marzouk, H. and Hadianfard, M. A. (2017), 'Monitoring of concrete shrinkage and creep using Fiber Bragg Grating sensors', *Construction and Building Materials* 137, 505–512.

Zajac, M., Skocek, J., Bullerjahn, F. and Ben Haha, M. (2016), 'Effect of retarders on the early hydration of calcium-sulpho-aluminate (CSA) type cements', *Cement and Concrete Research* 84, 62–75.

Zdanowicz, K. and Hansen, M. (2019), Chemische Vorspannung von textilbewehrten Sichtbetonbauteilen mit Quellbeton. Abschlussbericht., Technical report, Forschungsinitiative Zukunft Bau des Bundesinstitutes fur Bau-, Stadt- und Raumforschung, Nr. SWD-10.08.18.7-16.37.

Zdanowicz, K., Kotynia, R. and Marx, S. (2019), 'Prestressing concrete members with fibre-reinforced polymer reinforcement: State of research', *Structural Concrete* 20(3), 872–885.

Zdanowicz, K. and Marx, S. (2018*a*), Shrinkage and Expansion Strains in Self-compacting Concrete: Comparison of Methods of Measurements, *in* '"High Tech Concrete: Where Technology and Engineering Meet"', pp. 1–6.

Zdanowicz, K. and Marx, S. (2018*b*), Thin concrete panels prestressed with carbon textile reinforcement: flexural testing, *in* '5th International fib Congress "Smarter - Better - Stronger", Melbourne', Melbourne, Australia.

Zdanowicz, K. and Marx, S. (2019*a*), Chemical prestress of concrete with carbon textile reinforcement: Theoretical and analytical approaches, *in* 'Proceedings of the fib Symposium 2019: Concrete - Innovations in Materials, Design and Structures. Krakow, Poland', pp. 259–265.

Zdanowicz, K. and Marx, S. (2019*b*), Quasi-kontinuierliche Dehnungsmessungen an chemisch vorgespannten Balken, *in* 'Beiträge zur 7. DAfStb-Jahrestagung mit 60. Forschungskolloquium. Hannover', pp. 196–204.

Zdanowicz, K., Schmidt, B., Hansen, M. and Marx, S. (2020), 'Biege- und verbundverhalten von chemisch vorgespannten textilbewehrten betonelementen', *Beton- und Stahlbetonbau* 115(12), 972–979.

Zdanowicz, K., Schmidt, B., Naraniecki, H. and Marx, S. (2019), Bond behaviour of chemically prestressed textile reinforced concrete, *in* 'IABSE Symposium 2019', Guimaraes, Portugal, pp. 1–7.